Collins **wild guide**

Weather

Storm Dunlop

HarperCollins*Publishers* Ltd.
77-85 Fulham Palace Road
London
W6 8JB

The Collins website address is: www.collins.co.uk

First published in 2004

---

10 09 08 07 06 05
10 9 8 7 6 5 4 3 2

A catalogue record for this book is available from the British Library.

ISBN 0 00 716072 0

Colour reproduction by Saxon, Great Britain
Printed and bound in Great Britain by Scotprint

CONTENTS

INTRODUCTION

Weather affects us all, and modern weather forecasting has become increasingly technical and sophisticated. It uses highly advanced satellites to monitor conditions around the world, and employs some of the most powerful and fast supercomputers in existence to produce the forecasts. Television weather forecasts are among the most popular items on TV, yet most viewers have difficulty understanding exactly what those forecasts mean for their local area, and also why (for example) it is so difficult to predict the onset or duration of rain. Professional forecasts are becoming increasingly accurate, but most people fail to appreciate this when the conditions that they imagine will occur do not materialize.

It is hoped that this book will help readers to a greater understanding of the weather and its phenomena, and how simple observations may help them to appreciate the local effects that can greatly affect the weather that they actually experience. Reading this book will, I hope, help you to decide whether the clouds hold the menace of rain or suggest that little change will occur.

But the weather not only affects us in nearly everything that we do, but it also presents an ever-changing spectacle and some fascinating and beautiful phenomena to observe. Unfortunately many of these go unnoticed by the vast majority of people, who cast a cursory glance at the sky and think 'clear' or 'clouds' and pay little more attention, until, perhaps, they are drenched by an unexpected downpour.

Yet the sheer range of effects, whether of different cloud forms or of optical phenomena is amazing. I hope that this book will serve as an introduction to some of the beauties, as well as the complexities, of weather and the sky.

Chichester, 2003

HOW TO USE THIS BOOK

This book will help you to recognize many different aspects of weather that may be observed by anyone interested in the sky. After some general tips on observing and photographing the sky (pp.6–9), it introduces the basic types of cloud and provides a framework for identifying the many different forms that may be seen, which might otherwise seem to be overwhelming in their variety (pp.10–23).

The many different types of cloud (including some uncommon ones) are then described in detail (pp.24–91), followed by an outline of the way in which particular clouds are formed, helping you to understand why, when, and where they occur (pp.90–115). Some less common phenomena are then described (pp.116–125).

The next section (pp.126–157) offers descriptions that will help to guide you in the identification of some of the many optical effects and phenomena that may be seen. This is followed (pp.158–171) by a discussion of the various colours, and light and shadow effects, that constantly change the appearance of the sky.

A short section (pp.172–185) describes what meteorologists term 'precipitation': rain, snow, dew, frost and all the other forms that water may assume in the atmosphere.

It helps in the understanding of both weather as a whole, and also of specific phenomena to have an idea of the mechanisms governing the global circulation of air and other factors that determine both weather and climate. These aspects are covered in a specific section (pp.186–221), which is followed (pp.222–237) by a discussion of severe weather, including thunderstorms, tornadoes and hurricanes.

Satellites play such a large part in modern meteorology that they are discussed and some images are shown in a short section (pp.238–245).

Some reasonable inferences may be drawn from simple observations of the clouds and other factors, so these are briefly summarized (pp.246–249).

The book concludes with a glossary (pp.250–253) and some suggestions for further reading (p.254).

Observing the sky

Although it seems self-evident, the best way of learning about the sky is to keep looking at it. In our busy modern world, few people do more than glance at the sky, and many phenomena go unnoticed. Spending some time studying the sky's appearance will soon lead to a better understanding of the different types of clouds and how they change; of the vast range of other phenomena that may be observed, and how these all relate to the weather.

If possible, try to determine the direction and approximate strength of the wind. Remember that this is best shown by the movement of low clouds, but will always be slightly different from that at the surface – regardless of obvious obstacles such as buildings or hills – because of surface friction and other reasons described later. If there are no clouds, then you need to be out in the open so that you can determine the surface wind, which, as we shall see, enables you to estimate the direction of the true wind.

It is extremely useful to be able to make rough estimates of angular distances on the sky, and this may be achieved by using the hand at arm's length as shown in the diagram. Reducing glare will help to see many faint phenomena, so try using sunglasses – the mirror type are particularly helpful – or view a reflection of the sky in a pool of water or a sheet of glass. A pair of small pieces of plastic polarizing material are ideal, because any degree of darkening may be obtained by rotating one relative to the other.

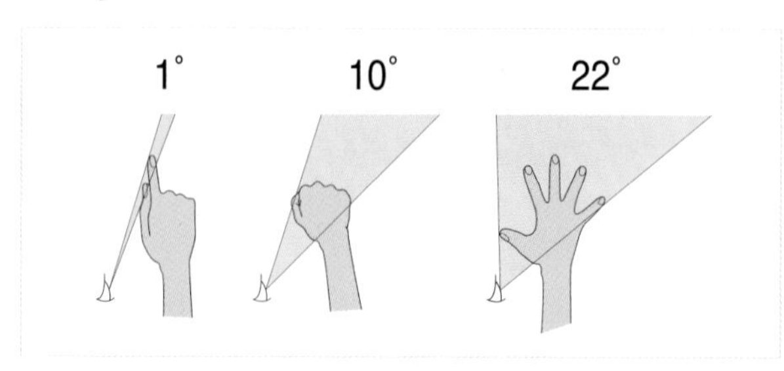

It is often helpful to use a pair of binoculars to examine clouds, making it easier to distinguish one type from another, especially at a distance. They will also help you to recognize the important changes that take place as a cumulus cloud becomes a cumulonimbus.

Finally, try to make a note of the date and time of any phenomena that you observe, particularly those that are rare, such as certain halo arcs.

Photographing the sky

Almost any camera (including digital ones) may be used successfully to photograph the sky, but a wide-angle lens or setting is often extremely useful. (It requires a 24 mm lens, or equivalent, to include the full extent of a secondary rainbow, for example.)

Most cameras with automatic exposure will tend to over-expose the sky, losing detail in bright areas. If the camera has spot metering, taking a reading from the brightest area and using the 'highlight' option generally guarantees good results. Otherwise, use exposure compensation to give slight under-exposure.

With print films the printing process is set to give the best rendering of foreground objects, and skies tend to lack detail. Slide films are more satisfactory in this respect, provided exposure has been correct in the first place. Digital images may, of course, be manipulated to increase contrast in the sky, but such changes should be used with care to avoid undue exaggeration of colours and contrasts.

It is good practice to use a skylight filter (such as a Kodak Wratten 1A) and a lens hood at all times. At high altitudes, a slightly pink filter (Wratten 2A or 2B) will help to avoid unnaturally blue skies.

A polarizing filter is extremely useful, because most objects in the sky (including the blue sky itself) are polarized to some degree. Different orientations of the polarizer will bring different objects into prominence and change the saturation of certain colours. Ensure, however, that you have the correct type of polarizer. Some automatic-exposure cameras require a circular polarizer, rather than the more usual linear ones.

Finally, make a habit of recording the date and time of any exposure. When photographing unusual phenomena it is also extremely important to record the focal length of the lens used (or the setting of a zoom lens), because this will enable the angular size of objects in the sky to be determined.

Clouds

At first sight identifying the different forms of clouds may seem almost impossible. No two clouds are the ever same and they appear to show a bewildering range of shapes and types. In fact, there are just ten main types that are relatively easy to recognize, so if you are just beginning to look at clouds it is simplest to concentrate on identifying those. Once you are fairly familiar with them, you can move on to identifying the various subsidiary forms.

Like many other natural objects, such as birds, flowers, or trees, clouds are classified into genera (types), species, and varieties, with some additional supplementary forms. Don't let this thought overawe you – once you know what to look for, the differences become quite apparent.

The various types of cloud may be grouped according to either their general form or their heights. Two forms are recognized by professional meteorologists: heaped (or cumuliform) clouds, and layer (stratiform) clouds. In general terms, these two forms occur, respectively, under unstable and stable conditions (explained shortly). A third group (cirriform), although not officially recognized, contains clouds that consist of ice crystals, and is helpful when first learning the different cloud types.

Clouds may also be classified according to their height. Although cloud heights are difficult to estimate, the clouds themselves are normally sufficiently distinct to make this a useful way of discussing their features, and the detailed discussion of cloud types that follows later (pp.24–90) is organized on this basis.

Clouds and the atmosphere

In the atmosphere, pressure may vary from place to place at the surface, but it always decreases with height. A parcel of air that ascends moves into a region of lesser pressure where it expands and cools. The opposite happens to a parcel of air that descends: it is compressed and warms. Because of the decrease in pressure with height, in the absence of other factors, the temperature would also decrease with height. In the lowest layer of the atmosphere, the troposphere, the overall decline is about 6°C per kilometre. This lapse rate is not constant, however, because the lower atmosphere normally contains layers of different temperatures and humidities.

A parcel of dry air – i.e., which may contain moisture, but where condensation into cloud droplets has not yet occurred – cools at about 10°C per kilometre. Once condensation has set in, a parcel cools at a slower rate of about 4.5–5°C per kilometre. If a parcel of air that ascends (for whatever reason) has a temperature higher than that of its surroundings, it will continue to rise. This condition is known as instability, and is a primary cause of the growth of clouds.

If, however, a parcel of air has a temperature below that of its surroundings, it tends to sink. Such conditions are said to be stable. Stability often arises because, instead of declining with altitude, the temperature in a particular layer rises. Such a layer is known as an inversion, and tends to prevent the growth of clouds, although if convection is very vigorous, it may be able to break through an inversion.

A major inversion, known as the tropopause, divides the troposphere (where most weather systems occur) from the overlying stratosphere. The tropopause generally limits the upwards growth of the largest cumulonimbus clouds. Its height varies from about 14–18 km over equatorial regions, to 5–8 km over the poles (where it may be indistinct in winter). The tropopause is not continuous, but exhibits breaks in its level, and these normally occur where there are major temperature contrasts between different air masses, particularly at the polar fronts (see p.187).

In the lower stratosphere to about 20 km, there is little temperature change with height, but it then begins to rise (because of absorption of solar radiation by the ozone) up to the stratopause at about 50 km. Above this altitude, in the mesosphere, the temperature again falls with height. Some clouds do form in the stratosphere, mainly cirrus (p.58) and the beautiful nacreous clouds (p.116). Noctilucent clouds (p.118) are found at the top of the mesosphere.

Cloud forms

Clouds that are produced by convection and occur as heaps or masses are described as cumuliform. (The word 'cumulus' is Latin for 'heap'). They are:

- cumulus
- cumulonimbus
- stratocumulus
- altocumulus
- cirrocumulus

The upper surfaces of all of these cloud types exhibit generally rounded heads or more obvious turrets, which are signs of instability. In the first two (cumulus and cumulonimbus), convection is often extremely active, particularly in cumulonimbus clouds, which may be

the source of a whole range of severe weather.
Clouds which occur under stable conditions, and where convection is absent occur as layers and are known as stratiform ('stratus' being Latin for 'layer'). These are:

- stratus
- nimbostratus
- altostratus
- cirrostratus

It may be noted that stratocumulus, altocumulus and cirrocumulus are often regarded as having both cumuliform and stratiform characteristics, because they occur in more-or-less extensive layers, but exhibit weak convection.

A third group of clouds, known as cirriform, is sometimes recognized ('cirrus' is Latin for 'curl', 'tuft', or 'wisp'). These are ice-crystal clouds and include:

- cirrus
- cirrocumulus
- cirrostratus

The term is also applied to the cirrus plumes associated with many cumulonimbus clouds.

Low clouds

The low clouds (cumulus, stratocumulus, and stratus) are generally taken to have bases between 300 m and 1500 m. Stratus (in particular) may extend down to the ground. The height of the base varies slightly from equator to poles, being somewhat higher at low latitudes (and during summer). However, the typical altitude range is 500–2,000 m.

Cumulus (Cu)

Rounded heaps of cloud at low levels
(p.24)

Stratus (St)

Essentially featureless, grey layer cloud at low level
(p.34)

Stratocumulus (Sc)

Heaps or rolls of cloud, with distinct gaps and heavy shading at low levels
(p.40)

Medium-level clouds

Three types of cloud (altocumulus, altostratus, and nimbostratus) are classified as being found at medium levels, although one (nimbostratus) often extends down almost to the ground. The height of the base varies much more with latitude than with the low clouds and is typically:

Tropics:	2,000–8,000 m
Middle latitudes:	2,000–7,000 m
High latitudes:	3,000–4,000 m

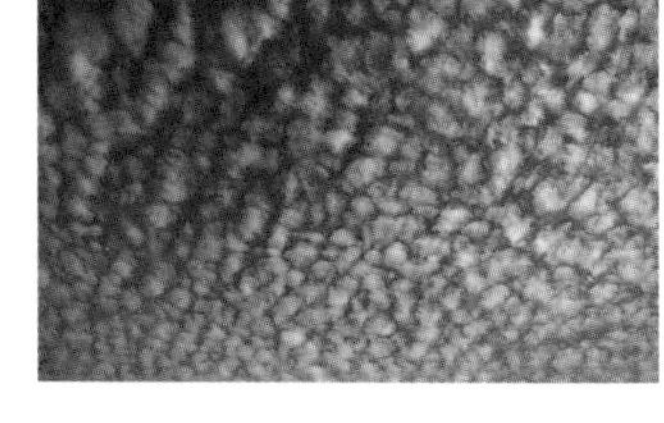

Altocumulus (Ac)

Heaps or rolls of cloud, showing distinct shading, and with clear gaps between them, in a layer at middle levels (p.46)

Altostratus (As)

Sheet of featureless, white or grey cloud at middle levels (p.52)

Nimbostratus (Ns)

Dark grey cloud at middle levels, frequently extending down towards surface, and giving prolonged precipitation (p.56)

High clouds

The three high clouds (cirrus, cirrocumulus, and cirrostratus) are generally defined as having bases at about 6 km or above (about 20,000 feet). However, as with the low and medium-level clouds, the bases may be lower at middle and high latitudes, and also during winter. Typical heights are:

Tropics:	6,000–18,000 m
Middle latitudes:	5,000–13,000 m
High latitudes:	3,000–8,000 m

Cirrus (Ci)

Fibrous wisps of cloud at high levels
(p.58)

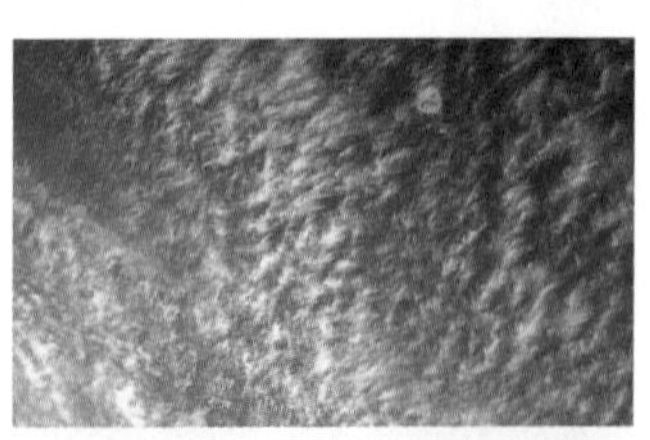

Cirrocumulus (Cc)

Tiny heaps of cloud with no shading, with clear gaps, in a layer at high levels
(p.66)

Cirrostratus (Cs)

Essentially featureless sheet of thin cloud at high levels
(p.70)

Clouds extending through more than one level

One cloud type (cumulonimbus) commonly extends from a low level up into the atmosphere, very frequently through the middle level and to heights normally associated with the cirrus family of clouds. Indeed, cumulonimbus clouds often reach throughout the troposphere – the layer closest to the ground in which the majority of weather phenomena occur.

Cumulonimbus (Cb)

Large towering cloud extending to great heights, with ragged base and heavy precipitation (p.74)

Cloud species

Differences in cloud shape and structure are described by the use of fourteen terms (species). These have three-letter abbreviations, and in this table the page numbers indicate those species that are illustrated elsewhere in the book. The cloud types that are not specifically illustrated – generally higher or lower clouds – show similar characteristics.

Species	Abbr.	Description	Genera
calvus	cal	Tops of rising cells lose their hard appearance and become smooth	Cb (p.76)
capillatus	cap	Tops of rising cells become distinctly fibrous or striated; obvious cirrus may appear	Cb (p.77)
castellanus	cas	Distinct turrets rising from an extended base or line of cloud	Sc, Ac (p.48), Cc, Ci
congestus	con	Great vertical extent; obviously growing vigorously, with hard, 'cauliflower-like' tops	Cu (p.32)
fibratus	fib	Fibrous appearance, normally straight or uniformly curved; no distinct hooks	Ci (p.60), Cs (p.72)
floccus	flo	Individual tufts of cloud, with ragged bases, sometimes with distinct virga	Ac (p.49), Cc, Ci
fractus	fra	Broken cloud with ragged edges and base	Cu (p.26), St (p.36)
humilis	hum	Cloud of restricted vertical extent; length much greater than height	Cu (p.28)

lenticularis	len	Lens- or almond-shaped clouds, stationary in the sky	Sc, Ac (p.50), Cc
mediocris	med	Cloud of moderate vertical extent, growing upwards	Cu (p.30)
nebulosus	neb	Featureless sheet of cloud, with no structure	St (p.37), Cs
spissatus	spi	Dense cloud, appearing grey when viewed towards the Sun	Ci (p.62)
stratiformis	str	Cloud in an extensive sheet or layer	Sc, Ac (p.51), Cc
uncinus	unc	Distinctly hooked, often without a visible generating head	Ci (p.63)

Cloud varieties

Cloud varieties describe variations in cloud transparency and the arrangement of cloud elements. The nine terms have standard two-letter abbreviations. It should be noted, however, that the features of more than one variety are often present simultaneously.

Variety	Abbr.	Description	Genera
duplicatus	du	Two or more layers	Sc (p.44), Ac, As, Cc, Cs
intortus	in	Irregularly tangled	Ci (p.64)
lacunosus	la	Curved thin cloud with regularly spaced holes, appearing like a net	Ac, Cc (p.69), Sc
opacus	op	Thick cloud that completely hides the Sun or Moon	St (p.38), Sc, Ac, As
perlucidus	pe	Extensive layer with gaps, through which blue sky, the Sun or Moon are visible	Sc, Ac
radiatus	ra	Appearing to radiate from one point in the sky	Cu (p.98), Sc, Ac, Ci As (p. 55)
translucidus	tr	Translucent cloud, through which the position of the Sun or Moon is readily visible	St, Sc, Ac, As (p.54)
undulatus	un	Layer or patch of cloud with distinct undulations	St (p.39), Sc (p.45), Ac, As, Cc, Cs
vertebratus	ve	Lines of cloud looking like ribs, vertebrae or fish bones	Ci

Accessory clouds

There are also three forms (known as accessory clouds) that occur only in conjunction with one of the ten main genera. These have standard three-letter abbreviations.

Feature	Abbr.	Description	Genera
pannus	pan	Ragged shreds of beneath main cloud mass	Cu, Cb, As, Ns (p.78)
pileus	pil	Hood or cap of cloud above rising cell	Cu, Cb (p.79)
velum	vel	Thin, extensive sheet of cloud, through which the most vigorous cells may penetrate	Cu, Cb (p.80)

Supplementary features

In addition, some specific forms (known as supplementary features) may be found in particular genera or species. These have standard three-letter abbreviations.

Feature	Abbr.	Description	Genera
arcus	arc	Arch or roll of cloud	Cb, Cu (p.81)
incus	inc	Anvil cloud	Cb (p.82)
mamma	mam	Bulges or pouches beneath higher cloud	Cb, Ci, Cc, Ac, As, Sc (p.84)
praecipitatio	pra	Precipitation that reachesthe surface	Cb, Cu, Ns (p.85)
tuba	tub	Funnel cloud of any type	Cb, Cu (p.86)
virga	vir	Fallstreaks: trails of precipitation that do not reach the surface	Ac, As, Cc, Cb, Cu, Ns, Sc, (Ci) (p.88)

Cumulus

Abbreviation: Cu

ID FACT FILE

APPEARANCE:
Individual heaps or 'pancakes' of cloud, generally white on top with flat, darker bases.

OPTICAL PHENOMENA:
Crepuscular rays.

APPARENT SIZE:
Very variable, depending on the cloud's distance from the observer. Horizontal extent frequently exceeds 5°.

HEIGHT OF BASE:
Below 2 km (approx. 6,500 ft).

PRECIPITATION:
Generally none, but may be significant with one species – cumulus congestus.

SPECIES:
fractus (p.26); humilis (p.28); mediocris (p.30); congestus (p.32)

VARIETY:
radiatus (p.98)

Cumulus clouds are very common – so common, in fact, that they are the type that most people imagine when they think of clouds. They are also one of the few cloud types to have gained a popular name, being frequently known as 'fair-weather clouds', and even meteorologists sometimes refer to them as 'fair-weather' cumulus. As these names imply, they are generally associated with fine weather, although one species, cumulus congestus, may sometimes produce significant amounts of rain. They are one of the three cloud types (with stratus and stratocumulus) that occur as low clouds.

Cumulus clouds are easy to recognize, being generally in the form of fluffy heaps of cloud with distinctly rounded tops and flat, darker bases. It is usually obvious that these flat bases are all at the same level. Only in their very early stages of formation or in the last stage of decay do they show ragged bases (in the species known as cumulus fractus). Depending on their depth relative to their horizontal extent, they are classified into the species cumulus humilis, cumulus mediocris, and cumulus congestus. There is just one cumulus variety, 'cumulus radiatus' and this is described later.

SEE ALSO:
Altocumulus (p.46); cumulus radiatus (p.98); stratocumulus (p.40); stratus (p.34)

Cumulus clouds are generally well separated from one another when they first form, although later in the day they may spread out to cover a considerable portion of the sky, and even give rise to layers of stratocumulus or altocumulus. Because they form from rising thermals of warm air, their flat bases clearly indicate the condensation level.

As with most clouds, the colour of cumulus depends strongly on the direction in which they are illuminated and the position of the observer. In full sunlight their tops may be brilliantly white, but when seen against the sun, they appear grey, and if sufficiently dense, almost black.

Cumulus fractus – broken cumulus

Abbreviation: Cu fra

ID FACT FILE

APPEARANCE: Ragged, individual wisps of cloud, often little more than misty patches in the air. Pale in colour, but appearing grey against the sun.

OPTICAL PHENOMENA: Generally none, very rarely crepuscular rays.

APPARENT SIZE: Very variable, but usually very small.

Height of base: Below 2 km (approx. 6,500 ft).

PRECIPITATION: None.

SEE ALSO: *Cumulus humilis (p.28); stratus fractus (p.36); pannus (p.78)*

Cumulus fractus are ragged fragments of low cloud that continuously and rapidly change their shape. These clouds are commonly seen early in the day, as heating of the ground begins, and small, weak thermals rise as far as the condensation layer. Initially, they are often little more than misty patches without well-defined edges. As the thermals become stronger and larger, distinct wisps of cloud appear. Many show slightly domed tops – a sign of the underlying shape of the parent thermals. If heating of the surface persists, these small clouds may eventually grow into another species, cumulus humilis.

Similar clouds often occur as cumulus decays, either because the parent thermals have become too weak to support growth – perhaps

because the clouds have moved away from the source of heat – or late in the day, when heating of the surface by the Sun has died down, causing any cumulus clouds to decay.

Cumulus fractus may be confused with stratus fractus, but cumulus fractus only arises when there is active convection (or when it is dying down), not under the quiet conditions that produce stratus fractus. Cumulus fractus also bears a close resemblance to the accessory cloud known as pannus, which may be found in association with cumulonimbus and nimbostratus (in particular) as well as cumulus congestus and altostratus. If none of these other types is present, however, then the ragged clouds can only be cumulus fractus or stratus fractus.

Cumulus humilis – flattened cumulus

Abbreviation: Cu hum

ID FACT FILE

APPEARANCE: Flattened cumulus that are much wider than their depth. White tops with darker bases. May be confused with Stratocumulus, when a layer of cumulus humilis is too far away for the gaps between the clouds to be readily visible.

OPTICAL PHENOMENA: Generally none, occasionally produce crepuscular rays.

APPARENT SIZE: Very variable, but usually more than 5° across.

HEIGHT OF BASE: Below 2 km (approx. 6,500 ft).

PRECIPITATION: None.

SEE ALSO: *Cumulus fractus (p.26); cumulus mediocris (p.30); stratocumulus (p.40); anticyclones (p.210)*

Cumulus humilis are very shallow clouds, with a vertical depth that is noticeably much less than their horizontal extent. Such cumulus commonly occur fairly early in the day, developing from initial cumulus fractus, several of which may merge to produce an individual cumulus humilis cloud. Cumulus humilis are generally much denser, however, so they reflect more light, and their tops are often brilliantly white, and the bases are distinctly darker. When seen against the light, of course, the dense clouds appear dark grey.

The cloud tops are usually slightly rounded, which is an indication that convection is still occurring within them. Under certain

conditions (particularly ahead of a warm front or when there is anticyclonic weather), the tops of the clouds may seem distinctly flattened, because their upward growth is restricted. The subsiding or encroaching air creates a stable layer (an inversion) which lies immediately above the visible cloud-tops and inhibits convection.

The spaces between the individual clouds are fairly large, unlike the gaps found in stratocumulus, with which this species might otherwise be confused. Cumulus humilis may, however, spread out horizontally as they develop, and turn into a layer of stratocumulus. If active convection continues they are likely to grow vertically into the species known as cumulus mediocris.

Cumulus mediocris – medium cumulus

Abbreviation: Cu med

ID FACT FILE

APPEARANCE: Cumulus showing upward growth. White with darker bases.

OPTICAL PHENOMENA: Generally none, occasionally produce crepuscular rays.

APPARENT SIZE: Variable, but usually more than 5° across.

HEIGHT OF BASE: Below 2 km (approx. 6,500 ft).

PRECIPITATION: None.

SEE ALSO: *Cumulus humilis (p.28); cumulus congestus (p.32); stratocumulus (p.40)*

The next stage in the growth of cumulus clouds is shown by the species known as cumulus mediocris. These are cumulus clouds that show signs of active growth, and that are often approximately triangular in profile, with a depth that is less than or similar to their horizontal extent. These clouds are dense, so the tops are white (or dark grey when seen against the light) and the bases are dark. The upper surfaces usually show distinct heads where convection and upward growth is continuing.

If there is wind-shear (a considerable change, usually an increase, in wind-speed with height), the growing towers often lean downwind. Cumulus humilis seen early in the day frequently develop into cumulus mediocris if the ground continues to be heated by

sunlight. In turn, cumulus mediocris may tend to spread horizontally, creating larger masses of cloud, or even becoming a layer of stratocumulus. With vigorous convection they may also grow vertically to become the species known as cumulus congestus.

Although cloud heights and sizes are notoriously difficult to estimate, the depth of cumulus mediocris clouds sometimes provides a useful guide. Quite frequently they are about one kilometre deep. Although this should be taken as only a very approximate figure, it may also be used to make a rough estimate of the cloud's horizontal extent.

Cumulus congestus – heaped cumulus

Abbreviation: Cu con

ID FACT FILE

APPEARANCE: Cumulus clouds with much greater vertical than horizont al extent. Tops brilliantly white, with darker bases.

MAY BE CONFUSED WITH: Cumulonimbus, when too distant for the distinguishing features to be seen.

OPTICAL PHENOMENA: Frequently cause crepuscular rays.

APPARENT SIZE: Variable, often more than 5–10° across when nearby.

HEIGHT OF BASE: Below 2 km (approx. 6,500 ft).

PRECIPITATION: Rain, particularly in tropical regions.

SEE ALSO: *cumulus mediocris (p.30); cumulonimbus (p.74)*

Cumulus congestus (often known as 'towering cumulus') are actively growing clouds with a greater vertical than horizontal extent. This species often develops from cumulus mediocris. In tropical regions they may be extremely deep at any time of year. This depth is so great that sufficient collisions can occur between water droplets to give rise to substantial rainfall, and cumulus congestus are a principal source of heavy rainfall in such regions. At temperate latitudes, cloud depths are more restricted for most of the year so cumulus congestus produce little rain. In summer, however, deep clouds can occur and give rise to heavy showers.

As with cumulus humilis and mediocris, the clouds are dense, appearing brilliantly white in sunlight, with darker bases. These bases are normally flat unless convection has ceased or rain is falling, when they may be ragged. Because of their depth, the tall cloud towers frequently lean downwind. Smooth tops to the clouds are an indication that convection is still occurring. At times it may be difficult to tell the difference between cumulus congestus and cumulonimbus clouds, but the latter's distinguishing characteristics are described shortly.

Stratus

Abbreviation: St

ID FACT FILE

APPEARANCE:
A low, almost featureless layer of cloud, with no large, or regular pattern of breaks.

OPTICAL PHENOMENA:
Occasionally coronae (p.128); very rarely haloes (p.132).

HEIGHT OF BASE:
Generally below 500 m

PRECIPITATION:
Infrequent, occasionally produces light drizzle or (very rarely) ice crystals at low temperatures.

SPECIES:
fractus (p.36)
nebulosus (p.37)

VARIETIES:
opacus (p.38);
undulatus (p.39)

SEE ALSO:
corona (p.128); haloes (p.132)

Stratus consists of a low, grey layer cloud with few distinct features. It is the lowest of all clouds with a base rarely higher than 500 m. As a result the cloud often hides the tops of hills and buildings. The base frequently descends to ground level, when it become obvious that there is no difference between stratus cloud and mist or fog. Because there is little precipitation, the base is not ragged but tends to be 'soft' and poorly defined. Being such a low cloud, stratus will sometimes fill valleys and leave hill and mountain tops in clear air above the layer.

Stratus generally consists of water droplets which may occasionally give rise to some optical phenomena. The most common is a corona, but when temperatures are very low, ice crystals may form and produce halo effects, although these are extremely rare.

Stratus forms under stable conditions and by its very nature shows few features. There are just two species: stratus fractus and stratus nebulosus, and three varieties. The latter are stratus opacus, where the cloud is so dense that it hides the Sun and Moon, and stratus translucidus where, conversely, the positions of these bodies are clearly visible. The other variety, stratus undulatus, shows distinct undulations in its surface – usually on its base, but sometimes visible on the top when the layer is seen from above.

Stratus fractus – broken stratus

Abbreviation: St fra

ID FACT FILE

APPEARANCE: Ragged fragments of grey cloud.

OPTICAL PHENOMENA: None except for crepuscular rays (p.166).

HEIGHT OF BASE: Generally below 500 m

PRECIPITATION: Infrequent, may occasionally produce light drizzle.

SEE ALSO: *cumulus fractus (p.26); pannus (p.78)*

Although stratus sometimes forms as a more-or-less complete layer, it also begins as ragged patches of cloud, known as stratus fractus, that gradually increase in size and number, until they form an unbroken sheet. Similarly, a layer of stratus usually fragments into stratus fractus when it begins to dissipate, either as a result of increased wind speed and turbulence, or from daytime heating.

Superficially, stratus fractus resembles cumulus fractus, but the different conditions under which it occurs are usually quite sufficient to differentiate between the two cloud varieties.

Stratus nebulosus – featureless stratus

Abbreviation: St neb

ID FACT FILE

APPEARANCE:
Featureless layer of grey cloud.

OPTICAL PHENOMENA:
A corona may sometimes be visible through thin cloud.

HEIGHT OF BASE:
Generally below 500 m.

PRECIPITATION:
Infrequent, may occasionally produce light drizzle and occasionally small snowflakes or ice crystals at low temperatures.

SEE ALSO:
cirrostratus (p.70); corona (p.128)

Stratus is one of the two cloud types (the other being cirrostratus) that may occur as an almost completely featureless layer. Such stratus nebulosus is often remarkably extensive and of uniform density. That density, however, may vary from a thin layer that allows the position of the Sun or Moon to be seen, to a dense sheet that completely hides them. These varieties are known as stratus nebulosus translucidus (St neb tr – translucent stratus) and stratus nebulosus opacus (St neb op – opaque stratus), respectively. The layer may also show undulations and thus be classed as stratus nebulosus undulatus (St neb un).

Stratus opacus – opaque stratus

Abbreviation: St op

ID FACT FILE

APPEARANCE: Featureless layer of thick grey cloud, hiding any sign of the Sun or Moon.

OPTICAL PHENOMENA: None.

HEIGHT OF BASE: Generally below 500 m.

PRECIPITATION: Infrequent, may occasionally produce light drizzle and occasionally small snowflakes or ice crystals at low temperatures.

SEE ALSO: *altostratus (p.52)*

Stratus is frequently so thick that it is impossible to detect the disk of either the Sun or the Moon through it. At the most there may be a very slightly lighter area of cloud, but often even this is missing and the layer is truly opaque. Despite the fact that the layer is thick, precipitation remains very slight. The light drizzle or snow that sometimes falls from any form of stratus is often the result of the cloud being 'seeded' with ice crystals falling from a higher layer, such as altostratus.

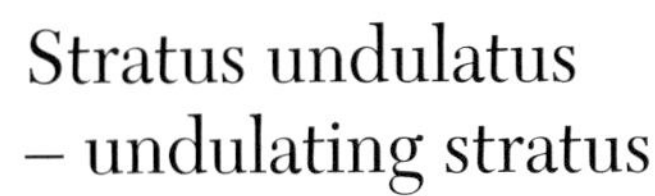

Stratus undulatus – undulating stratus

Abbreviation: St un

ID FACT FILE

APPEARANCE: Normally featureless with undulating base.

OPTICAL PHENOMENA: None reported.

HEIGHT OF BASE: Generally even the crests are below 500 m.

PRECIPITATION: As stratus, i.e., infrequent, most commonly drizzle.

SEE ALSO: *wave clouds (p.102)*

Generally, stratus occurs with a moderately well-defined base at a single level, although this is quite commonly diffuse. Sometimes, however, the base may show significant undulations, causing it to be termed stratus undulatus. This is particularly the case in hilly or mountainous country, even if the undulations may not warrant the cloud being defined as a wave cloud (described later). On very rare occasions, the stratus may exhibit a striated nature, as in the photograph shown here, but this is very unusual for this type of cloud.

It is, however, quite common for the upper surface of stratus to show undulations. This is most readily seen from an aircraft or from a mountain when looking down on a layer of stratus that might, from the ground, appear as valley fog.

Stratocumulus

Abbreviation: Sc

ID FACT FILE

APPEARANCE:
A layer of lumps, rolls or flattened cloud elements, always more than 5° across at 30° elevation, with clear sky between the elements.

OPTICAL PHENOMENA:
Very occasionally coronae (p.128) in thin patches; crepuscular rays (p.166).

HEIGHT OF BASE:
Generally around 500 m, but up to 2,000 m.

Stratocumulus displays both stratiform and cumuliform features. It is a very common type of cloud (being particularly frequent over the oceans) and consists of a layer of elements in the form of heaps, rolls, or flat 'pancakes' of cloud, with clear spaces between them.

The individual cloudlets show distinct shading and are always larger then 5° across, measured at an elevation of 30° above the horizon. The density and colour of the cloud varies greatly, from almost white to extremely dark grey. The gaps between the cloud elements are always present to a greater or lesser extent and clear sky may be seen through them when they are overhead. Sunlight shining through the spaces often gives rise to crepuscular rays.

PRECIPITATION: Occasionally light drizzle or, under cold conditions, light snow or ice crystals.

SPECIES: castellanus (p.42); lenticularis; stratiformis (p.43)

VARIETIES: duplicatus (p.44); lacunosus; opacus; perlucidus; radiatus; translucidus; undulatus (p.45)

SEE ALSO: *cumulus (p.24); inversion (p.12)*

This type of cloud may form through the break-up of an existing layer of stratus, often when cooling of the upper surface initiates convection within the layer. It also commonly arises when cumulus clouds reach an inversion and spread out sideways.

Stratocumulus gives little precipitation, but occasionally, especially when seeded by ice crystals falling from a higher layer (such as altostratus), may produce a light drizzle.

Stratocumulus may be confused with stratus when seen from a distance where the breaks are invisible, and also with higher altocumulus. The latter, however, has individual elements that are less than 5° across.

Stratocumulus castellanus – turreted stratocumulus

Abbreviation: Sc cas

ID FACT FILE

APPEARANCE: Separate cloud elements, but with turrets or towers rising from the upper surface.

HEIGHT OF BASE: As for stratocumulus: generally about 500 m, but sometimes as high as 2,000 m.

PRECIPITATION: Very slight drizzle is very occasionally observed.

SEE ALSO: *altocumulus castellanus (p.48); instability (p.12)*

Although stratocumulus normally forms under fairly quiet conditions with restricted convection, occasionally there is sufficient heating for moderate convection to begin within the cloud layer. This produces towers of cloud reaching upwards, and the cloud is then classed as stratocumulus castellanus.

This sign of instability is significant for gaining an idea of how the weather may develop. The latent heat in the layer is a potential source of energy, and if convection from the ground reaches the humid layer it may undergo explosive development, turning into cumulus congestus or cumulonimbus, sometimes initiating the formation of heavy showers.

Stratocumulus stratiformis – extensive stratocumulus

Abbreviation: Sc str

ID FACT FILE

APPEARANCE:
An extensive layer of stratocumulus covering the whole sky.

HEIGHT OF BASE:
Generally about 500 m, but sometimes as high as 2,000 m.

PRECIPITATION:
Very slight drizzle may sometimes occur.

Although, by its very nature, stratocumulus tends to occur in layers, sometimes it may be just a small patch of cloud, and at other times it may be extremely extensive and cover essentially the whole sky. It is then classed as stratocumulus stratiformis. Not only are the individual cloud elements large in diameter, but they are generally close together. Paradoxically, this situation may arise both when a sheet of stratus begins to break up, and also when stratocumulus that has built up during the day begins to merge into an unbroken sheet of stratus.

Stratocumulus duplicatus – multiple stratocumulus

Abbreviation: Sc du

ID FACT FILE

APPEARANCE: Patches or more extensive sheets of stratocumulus at more than one level.

HEIGHT OF BASE: Generally about 500 m, but sometimes as high as 2,000 m.

PRECIPITATION: Generally none, but slight drizzle on rare occasions.

Any of the layer clouds may occur at more than one level, separated by a clear layer (or layers) but this may be difficult to see from the ground, especially in the case of stratus, altostratus and cirrostratus. Even if there are breaks in the lower layer, it is often hard to be certain that a higher layer exists. With stratocumulus, however, it is normally possible to see that there is a second layer slightly higher in altitude, especially when the cloud elements of the lower layer are fairly widely spaced. Such stratocumulus duplicatus may occur as either small patches or more extensive layers at any one level.

Stratocumulus undulatus – stratocumulus billows

Abbreviation: Sc un

ID FACT FILE

APPEARANCE: Regularly spaced rolls of cloud in a stratocumulus layer.

HEIGHT OF BASE: Generally about 500 m, but sometimes as high as 2,000 m.

PRECIPITATION: Generally none.

SEE ALSO: *wave clouds (p.102)*

Quite frequently stratocumulus occurs as billows (stratocumulus undulatus), which sometimes form a complete layer. As with most billows, they generally lie approximately at right-angles to the wind prevailing at their level, in contrast to normal stratocumulus elements, which tend to be elongated downwind. However, there are departures from regularity in size, spacing and orientation of the billows.

An important point to notice is that stratocumulus billows (and all other billows) move downwind and, as with most clouds, tend to change their shape with time. They should not be confused with wave clouds, which remain stationary in the sky, and which may remain unchanged for hours on end.

Altocumulus

Abbreviation: Ac

Altocumulus is a medium-level cloud, which occurs as individual, rounded masses which always have areas of clear sky between them. It may appear in small, isolated patches, but is often part of an extensive layer and may sometimes cover the whole sky. Like stratocumulus, but unlike cirrocumulus, the individual cloud elements always show some darker shading. They are also larger than cirrocumulus cloud elements (being more than 1° across, measured 30° or more above the horizon), and smaller than stratocumulus cloud masses, which are always greater than 5° across.

Altocumulus clouds may contain either water droplets, which are usually supercooled, or ice crystals (or both). They may, therefore, exhibit a range of optical phenomena, depending on which form of water is predominant. Water-droplet effects (iridescence and coronae) are quite frequently seen (particularly around the edges of individual cloud elements, or where the cloud is thin). Very occasionally, altocumulus may exhibit phenomena caused by ice crystals, such as parhelia (mock suns) and sun pillars.

Altocumulus may be formed by different mechanisms, two of which are: the spreading of convective cloud (such as cumulus or cumulonimbus) at a middle level, and the breakdown of a sheet of altostratus. (The latter

ID FACT FILE

APPEARANCE: Heaps, rolls or 'pancakes' of cloud, with darker shading and clear sky between them.

OPTICAL PHENOMENA: Iridescence or corona sometimes visible, rarely mock suns or sun pillars.

APPARENT SIZE: Individual cloudlets are 1–5° across (larger than cirrocumulus, smaller than stratocumulus).

HEIGHT OF BASE: 2–6 km (approx. 6,500–20,000 ft).

PRECIPITATION: Rarely reaches the ground. Virga (p.88) frequently seen.

SPECIES: castellanus (p.48); floccus (p.49); lenticularis (p.50); stratiformis (p.51)

VARIETIES:
duplicatus; lacunosus; opacus; perlucidus; translucidus; undulatus

SEE ALSO:
altostratus (p.52); cirrocumulus (p.66) stratocumulus (p.40); iridescence (p.130); corona (p.128); parhelion (p.138); sun pillar (p.142)

often occurs under quiet conditions as the heat of the Sun increases during the day.) Active cumulus and cumulonimbus often raise large quantities of water vapour to intermediate es of altocumulus behind them as they move downwind.

Altocumulus clouds include many species and varieties and are some of the most beautiful clouds in the sky.

Altocumulus castellanus – turreted altocumulus

Abbreviation: Ac cas

ID FACT FILE

APPEARANCE:
Towers rising from altocumulus cloud, generally arranged in lines.

OCCURRENCE:
When strong convection has set in at middle levels. An indicator of probable vigorous convective activity to come.

SEE ALSO:
cumulus congestus (p.32); stratocumulus castellanus (p.42); cumulonimbus (p.74); thunderstorms (p.226)

Altocumulus clouds sometimes develop towers that rise above the general level of the cloud layer. Such altocumulus castellanus turrets often occur in lines, and as with the similar growing heads seen in stratocumulus castellanus, they are a sign of instability at that height. As such, they often precede outbreaks of thunderstorms, because when cumulus congestus or cumulonimbus clouds reach the unstable layer they undergo a surge in growth, which is often sufficient to convert them into fully fledged storms.

Altocumulus floccus – tufted altocumulus

Abbreviation: Ac flo

ID FACT FILE

APPEARANCE:
Isolated puffs of altocumulus cloud, often in great numbers.

OCCURRENCE:
An indication of instability at middle levels. Also shows that vigorous convective activity is likely to develop.

SEE ALSO:
altocumulus castellanus (opposite); cumulus congestus (p.32); cumulonimbus (p.74); thunderstorms (p.226)

Occasionally the sky appears to be covered in small lumps of middle-level cloud, looking just like tufts of cotton wool. The individual cloud elements are normally smaller than those seen in 'ordinary' altocumulus, and close examination shows that they have rounded tops – an indication of instability and convection at that altitude. As with altocumulus castellanus, the presence of this humid layer is an indication that there may be outbreaks of thundery showers later that day, or possibly the next day, if convection from the surface reaches the layer and gains enough energy to become vigorous cumulonimbus clouds.

Altocumulus lenticularis – lenticular altocumulus

Abbreviation: Ac len

ID FACT FILE

APPEARANCE:
Lens- or almond-shaped clouds or longer smooth clouds parallel to high ground.

OCCURRENCE:
With wind, steady in strength and direction, that encounters hills or mountains. This species is an indication of a stable, humid layer at height.

SEE ALSO:
wave clouds (p.102)

Altocumulus lenticularis are smooth almond- or lens-shaped wave clouds that hang essentially motionless in the sky. They may persist for a long time, provided the wind strength and direction remain fundamentally unchanging. Depending on the circumstances, they may be a single cloud hanging over a peak, or a train of clouds stretching downwind – sometimes even so far that the hills or mountains causing them are invisible. They may also occur as long lines of cloud parallel to the mountain range. Examination with binoculars will often show how they form at the upwind edge and evaporate downwind. They occasionally have tiny corrugations parallel to the wind direction.

Altocumulus stratiformis – extensive altocumulus

Abbreviation: Ac str

ID FACT FILE

APPEARANCE:
An extensive layer of individual elements of altocumulus covering essentially the whole sky.

OCCURRENCE:
Fairly common and an indication, unless instability develops, of little rapid change in the weather.

SEE ALSO:
altocumulus castellanus (p.48); altocumulus floccus (p.49)

As with stratocumulus, altocumulus often occurs in relatively isolated patches, but altocumulus also frequently appears as an extensive sheet of similar-sized cloudlets covering the whole sky. This it often an extremely beautiful sight, especially at sunrise and sunset, when lit by a low sun. Generally, however, the most striking phase is short-lived, because of the fairly rapidly changing illumination. Although this species is an indication of a widespread, humid layer, unless it develops into either altocumulus floccus or altocumulus castellanus, it does not herald any immediate change in the weather.

Altostratus

Abbreviation: As

ID FACT FILE

APPEARANCE:
Extensive layers (or smaller patches) of nearly featureless grey cloud.

OCCURRENCE:
Widespread ahead of warm and occluded frontal systems, and often accompanying cold fronts and active storm systems

SPECIES:
None

VARIETIES:
Duplicatus; opacus; translucidus (p.54); radiatus (p.55); undulatus

SEE ALSO:
Cirrostratus (p.70); depressions (p.214); fronts (p.204); nimbostratus (p.56); stratus fractus (p.36)

Altostratus is a relatively featureless middle-level layer cloud, which may occur both as fairly small patches and as extremely widespread sheets, covering areas hundreds of kilometres across. It shows little variation in structure, apart from its thickness, which may vary from a very thin layer, through which the outline of the Sun or Moon may be seen, to heavy sheets thousands of metres thick.

It consists of a mixture of water droplets and ice crystals or snowflakes, and is generally light to dark grey, occasionally showing a fibrous character. Its most characteristic feature is that the Sun and Moon appear as if through frosted glass. Neither can be seen through thick altostratus, which produces a diffuse light, such that objects on the ground do not cast shadows.

Extensive sheets of altostratus occur at the warm or occluded fronts of a depression, usually preceded by cirrostratus, and itself becoming the lower, denser nimbostratus. In addition, the activity at cold fronts and large cumulonimbus often lift humid air to middle layers, producing altostratus that trails behind the active centres.

Frontal altostratus often produces a large amount of persistent rain or snow, that may reach the surface. It is frequently accompanied by prominent virga, and even when the

precipitation evaporates before reaching the ground, it may cause cooling of the lowest layer and produce stratus fractus (pannus).

By contrast, the layers of altostratus that are created by individual active cumulonimbus or by more extensive multicell or supercell systems (p.226), are relatively quiet and do not normally produce much precipitation. A layer of altostratus produced when cumulus spreads out beneath an inversion (often initially as altocumulus) similarly give rise to little precipitation.

Altostratus translucidus – transparent altostratus

Abbreviation: As tr

ID FACT FILE

APPEARANCE:
Thin altostratus that allows the outline of the Sun or Moon to be seen.

OCCURRENCE:
A common stage, both as the cloud thickens (into altocumulus opacus), and as it disperses, perhaps after the passage of a front.

SEE ALSO:
corona (p.128); iridescence (p.130); fronts (p.204)

Quite frequently, altostratus is thin enough for the outline of the disk of the Sun or Moon to be seen, although this is always indistinct. It is common for the cloud to thicken and the Sun to disappear, when the cloud would be regarded as having become altostratus opacus. Because altostratus consists of a mixture of water droplets and ice crystals, it rarely exhibits any optical phenomena. Occasionally the edges of a sheet of altostratus may be thin enough for iridescence to occur or a portion of a corona to appear.

Altostratus radiatus – altostratus in parallel lines

Abbreviation: As ra

ID FACT FILE

APPEARANCE:
Streaks of cloud aligned with the wind direction, rarely with clear sky showing between them.

OCCURRENCE:
Quite commonly seen as altostratus thickens ahead of a warm front.

SEE ALSO:
cirrus radiatus (p.65); cumulus radiatus (p.98); wave clouds (p.102)

From time to time altostratus is seen in long streaks stretching parallel to the wind direction. These seem to converge in the distance, both upwind and downwind, so the cloud is known as altostratus radiatus. Because there is rarely clear sky between the streaks, the cloud does not appear as dramatic as cumulus or cirrus radiatus. The streaks themselves are narrower than the undulations that also occur in altostratus, perpendicular to the wind direction. The latter form (altostratus undulatus) is normally a wave cloud.

Nimbostratus

Abbreviation: Ns

ID FACT FILE

APPEARANCE:
Heavy grey cloud with a diffuse base, often with separate pannus beneath it and producing large amounts of precipitation.

OCCURRENCE:
A common cloud at all frontal systems.

SPECIES:
None

VARIETIES:
None

SEE ALSO:
altostratus (p.52); cumulus fractus (p.26); fronts (p.204); stratus fractus (p.36)

Although nimbostratus is classed as a middle-level cloud, its base frequently extends more-or-less to the surface, even over low-lying ground. It is a grey or dark grey cloud, generally with a diffuse base, that produces large quantities of rain or snow. It commonly develops from altostratus at the approach of a warm front, where the cloud gradually thickens and lowers to become nimbostratus. It is also common at both cold and occluded fronts, and at the latter contributes most of the precipitation. It is always thick and completely hides the Sun. The rain falling from nimbostratus frequently cools the air below it to such an extent that cumulus fractus and stratus fractus (pannus) form in the saturated air beneath it.

Nimbostratus may also arise when altocumulus or stratocumulus thickens considerably, but this is relatively rare. Just as large cumulonimbus clouds or storm systems may create altostratus, they can also spread out to produce nimbostratus, which may give precipitation for some time after the main system has passed. Naturally, the area affected is much smaller than the vast regions covered by nimbostratus in depressions.

The cloud shown is slightly atypical, in that the photograph was taken during a slight break in the cloud at a warm front. The cumuliform

cloud in the distance was rapidly replaced by an unbroken sheet of nimbostratus (p.175) with extensive pannus (seen in the foreground).

Nimbostratus itself varies in composition. It may consist entirely of water droplets (cloud droplets and larger raindrops) or of snowflakes and ice crystals. Frequently both water droplets and ice crystals are present at the same level.

The precipitation from frontal nimbostratus is commonly arranged in bands, roughly parallel with the line of the front, separated by lines of lesser or no precipitation. The nimbostratus at slow-moving occluded fronts may produce days of rain or snow, and has frequently been the cause of major floods and deep snow-fall.

Cirrus

Abbreviation: Ci

Cirrus is an ice-crystal cloud found at high altitudes – indeed some cirrus clouds may occasionally occur in the lowest region of the stratosphere. They generally consist of elongated wisps or trails of cloud, although sometimes very dense patches occur. They are the most commonly recognized of the high clouds, especially the elongated tufts known popularly as 'mares' tails'.

Cirrus clouds are often associated with periods of fine, anticyclonic (high-pressure) weather, as well as with the more changeable weather found in depressions. In the form of cirrus plumes this cloud type also accompanies major storm systems, and in the case of tropical cyclones, forms a vast cirrus shield high above the turbulent weather beneath.

Because cirrus clouds consist of ice crystals, they occasionally display various optical effects. The most frequent are parhelia (mock suns), which may appear particularly brilliant. Generally, the areas of cirrus are too small for many of the halo arcs to appear, but portions have sometimes been seen.

ID FACT FILE

APPEARANCE:
Wisps of high cloud, generally white, but may appear grey when very dense.

OCCURRENCE:
Under a wide range of meteorological conditions.

SPECIES:
castellanus; fibratus (p.60); floccus (p.61); spissatus (p.62), uncinus (p.63)

VARIETIES:
duplicatus; intortus (p.64); radiatus (p.65); vertebratus

SEE ALSO:
depressions (p.214); parhelion (p.138); storms (p.226); tropical cyclones (p.236)

Cirrus fibratus

Abbreviation: Ci fib

ID FACT FILE

APPEARANCE: Moderately straight or slightly wavy streamers of cloud, without hooks, and aligned approximately parallel to one another.

OCCURRENCE: Under a wide range of meteorological conditions.

SEE ALSO: *cirrus intortus (p.64); cirrostratus (p.70); haloes (p.132); parhelion (p.138)*

Cirrus often appears in the form of fairly straight trails or streamers, without any of the distinct tufts of cloud (generating heads) where the ice crystals are actually forming. Although there may be some slight waviness in the trails, they do not show any of the distinct hooks that are seen in cirrus uncinus, for example, and are generally parallel to one another, unlike the tangled wisps of cirrus intortus.

It is sometimes difficult to decide whether a cloud is cirrus fibratus or cirrostratus fibratus, because both show a somewhat similar fibrous character. Generally, however the sky between the streamers of cirrus fibratus is clear, whereas with cirrostratus fibratus there is usually a distinct veil of cloud over a large area of sky. In addition, cirrostratus – at least when it is fairly thin – usually displays various halo phenomena, in particular the 22° and 46° haloes and parhelia.

Cirrus floccus

Abbreviation: Ci flo

ID FACT FILE

APPEARANCE:
Large numbers of wispy tufts of cloud, sometimes with distinct short virga.

OCCURRENCE:
When there is instability at cirrus levels with active creation of ice crystals.

SEE ALSO:
cirrostratus (p.70); cirrus fibratus (opposite); cirrus uncinus (p.63); virga (p.83)

Cirrus occasionally assumes the form of distinct wispy tufts of cloud with ragged bases. These are generating heads where the ice crystals are forming. Frequently they are accompanied by virga that hang beneath them and which stream away behind the clouds. Generally, however, these trails are fairly short. (Otherwise the cloud would probably be classified as cirrus uncinus.) As with the lower cloud form of altocumulus floccus, these tufts of cloud are an indication of instability at height. Unlike the humid layer in altocumulus, the instability at cirrus level does not add significantly to the activity of cumuliform clouds that reach it.

Cirrus spissatus – dense cirrus

Abbreviation: Ci spi

ID FACT FILE

APPEARANCE: Dense clumps of cirrus that appear dark grey against the light and hide the Sun or Moon.

OPTICAL PHENOMENA: Bright parhelia may occur in the thinner cloud at the edge of a patch of cirrus spissatus.

SEE ALSO: *cirrus fibratus (p.60); cirrus uncinus (opposite); cumulonimbus incus (p.82); parhelion (p.138)*

Cirrus frequently appears in very dense clumps, known as cirrus spissatus, that are so dense that they block the light of the Sun (or Moon) and appear dark grey when seen against the light. This is in complete contrast to the thin wisps of cirrus fibratus or cirrus uncinus. This particular cloud species (spissatus) is unique to cirrus, but is quite common, because it frequently occurs as the cirrus plumes or anvils cumulonimbus clouds (cumulonimbus incus).

Although not normally associated with any optical phenomena, because the cloud is very thick, sometimes the edges of a patch of cirrus spissatus are thin enough for bright parhelia (mock suns) to appear.

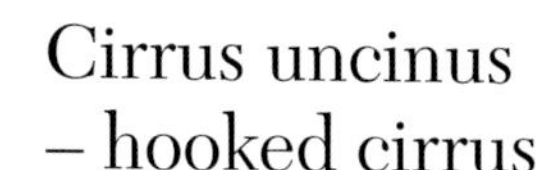

Cirrus uncinus – hooked cirrus

Abbreviation: Ci unc

ID FACT FILE

APPEARANCE:
Streamers of ice crystals with a distinct hook at one end.

OPTICAL PHENOMENA:
None.

SEE ALSO:
cirrocumulus (p.66); cirrus floccus (p.61); virga (p.88)

Another cloud species unique to cirrus is cirrus uncinus. This consists of long streamers of ice crystals that have a distinct hook at the upwind end. Although in many cases there is a slightly thicker cloud at the tip of the streamer, this generating head does not have a rounded appearance. (If it did, the cloud would be regarded a cirrocumulus with virga or possibly cirrus floccus.) In many cases, however, no obvious generating head is seen.

The angle of the streamers of ice crystals gives an indication of the wind speed at the clouds' altitude. The stronger the wind, the closer the streamers are to the horizontal. On very rare occasions, extremely long and nearly vertical trails have been observed, indicating a deep layer of nearly still air. The clouds could no longer be regarded as cirrus uncinus.

Cirrus intortus – tangled cirrus

Abbreviation: Ci in

ID FACT FILE

APPEARANCE:
Apparently randomly entangle streamers of ice crystals, with no obvious pattern.

OPTICAL PHENOMENA:
None.

SEE ALSO:
cirrocumulus (p.66); cirrus floccus (p.61); virga (p.88)

Frequently cirrus assumes the form of dense tangles of cloud, many of which are so thick that they might also be classified as cirrus spissatus. The strands appear to be entangled at random, with no obvious pattern. Neither is there any recognizable pattern to the distribution of the individual clumps of cloud in the sky. This variety is sometimes found behind frontal systems, and also associate with the quiet weather of anticyclonic conditions.

Cirrus radiatus – parallel streaks of cirrus

Abbreviation: Ci ra

ID FACT FILE

APPEARANCE:
Long streamers of cirrus that appear to converge towards a point on the horizon.

OCCURRENCE:
Particularly common in jet-stream cirrus.

SEE ALSO:
billows (p.100); jet streams (p.189); jet-stream clouds (p.114)

Cirrus may appear as long, parallel streamers of cloud that stretch far across the sky. The trails of ice crystals are so numerous and so long that, because of perspective, they appear to converge at the horizon. This variety, cirrus radiatus, is extremely common in jet-stream cirrus, where it is frequently accompanied by billows running perpendicular to the wind direction.

Cirrocumulus

Abbreviation: Cc

ID FACT FILE

APPEARANCE:
Small cloud elements blue-white or bluish in colour.

OCCURRENCE:
Through convection setting in to a shallow layer of cloud, or as wave clouds.

OPTICAL EFFECTS:
Occasionally coronae or iridescence.

SPECIES:
castellanus; floccus (p.68); lenticularis; stratiformis

VARIETIES:
undulatus; lacunosus (p.69)

SEE ALSO:
billows (p.100); corona (p.128); iridescence (p.130); and wave clouds (p.102)

Cirrocumulus consists of tiny cloud elements that are so high that they appear less than 1° across – measured, as usual, 30° up from the horizon. This differentiates them from high altocumulus, but an important additional distinguishing feature is that they rarely show any shading, appearing a fairly uniform pale blue or white in colour. The cloud elements are often difficult to see against the sky because they are thin, and thus low in contrast.

Cirrocumulus mainly consists of ice crystals and some supercooled water droplets. It often appears as a series of fairly regular cloud masses, ripples or billows. Such a regular pattern is commonly known as a 'mackerel sky', although the term is also applied to high altocumulus. Because the cloud is so thin, the Sun and Moon are always visible through it, and may produce certain optical effects.

The cloud particles are very even in size, so the most frequent effects are coronae and iridescence, rather than halo phenomena.

Frequently, the tufts of cloud have ragged bases, or trail short virga, but generally the precipitation is very slight and evaporates high in the atmosphere.

Cirrocumulus is normally caused by gentle convection occurring within a humid layer, which is usually started by the cloud radiating heat away to space. It also quite frequently appears as a wave cloud (cirrocumulus lenticularis).

Cirrocumulus floccus

Abbreviation: Cc flo

ID FACT FILE

APPEARANCE: Small cloud elements blue-white or bluish in colour.

OCCURRENCE: Through convection setting in to a shallow layer of cloud, or as wave clouds.

OPTICAL EFFECTS: Occasionally coronae or iridescence.

SEE ALSO: *altocumulus castellanus (p.48); altocumulus floccus (p.49)*

Cirrocumulus floccus consists of small tufts of cloud, with rounded heads and ragged bases. There are often distinct virga beneath these tufts, and these trails are generally short, and rarely very prominent. Just like altocumulus floccus, the cirrocumulus species may be present in large numbers all across the sky. Unlike the lower cloud, however, they rarely present such a striking appearance because of their pale colour and low contrast against the sky.

Like cirrocumulus castellanus, cirrocumulus floccus does indicate instability at high altitude. Any cumuliform cloud that reaches that level will gain some energy from the latent heat in the layer, but this is less than that gained from a lower humid layer containing altocumulus castellanus or floccus.

Cirrocumulus lacunosus

Abbreviation: Cc lac

ID FACT FILE

APPEARANCE:
Patches of cirrocumulus with clear holes.

OCCURRENCE:
Often occurs when a layer of cloud is dissipating.

SEE ALSO:
altocumulus (p.46); stratocumulus (p.40)

The cloud variety known as lacunosus is rarely seen, although it may occur in three cloud types: stratocumulus, altocumulus, and cirrocumulus. The official description in the World Meteorological Organisation's *International Cloud Atlas* says that it consists of a layer with more-or-less regular holes giving an appearance like a net or honeycomb. Frequently, however, the holes are very irregular.

Because of the cloud type's generally low contrast, cirrocumulus lacunosus often goes unnoticed, but small patches occur moderately frequently, usually surrounded by normal cirrocumulus. It usually appears as a layer of cirrocumulus begins to dissipate.

Cirrostratus

Abbreviation: Cs

ID FACT FILE

APPEARANCE:
A thin veil of cloud, which may thicken and show distinct striations.

OCCURRENCE:
A common precursor of warm fronts.

OPTICAL EFFECTS:
Numerous, including haloes, parhelia, circumzenithal arcs, and other halo effects.

SPECIES:
fibratus (p.72); nebulosus

VARIETIES:
duplicatus; undulatus (p.73)

SEE ALSO:
circumzenithal arcs (p.139); cirrocumulus (p.66); cirrus (p.58); haloes (p.132); parhelion (p.138)

Cirrostratus is a very common cloud, yet it frequently goes completely unnoticed. The reason for this apparent contradiction is that it generally starts as a very thin veil of high cloud, often with some striations, and gradually thickens. In its initial stages it hardly affects the light from the Sun, although later people become aware that the Sun has lost some of its warmth.

Cirrostratus is often found at warm fronts. Early wisps of cirrus become more numerous and gradually spread out into cirrostratus. It is at this early stage that this ice-crystal cloud may produce various halo effects, including 22° and 46° haloes, parhelia (mock suns), as well as brilliantly coloured circumzenithal arcs. Eventually, the cirrostratus thickens and the optical effects disappear. Gradually the sheet of cloud lowers until water droplets are able to exist, and it has become middle-level altostratus.

Sometimes the ice crystals that fall from cirrocumulus may give rise to a sheet of cirrostratus, and occasionally the cirrus plume of a cumulonimbus cloud may produce a similar layer. On less frequent occasions, cirrostratus may remain after a sheet of altostratus has decayed.

It has been estimated that at middle latitudes cirrostratus is visible about once every three days. Because many of the optical effects are short-lived, if you wish to see them, you need

to be alert for the presence of thin cirrostratus. A quick glance at the sky, covering the position of the Sun with the hand, will usually reveal any effects present. It is usually worthwhile to check at frequent intervals until the cirrostratus becomes thick and readily visible even without hiding the Sun.

Cirrostratus fibratus

Abbreviation: Cs fib

ID FACT FILE

APPEARANCE:
A fairly dense layer of fibrous cirrus.

OCCURRENCE:
Commonly occurs as one stage in the growth of cloud at warm fronts.

OPTICAL EFFECTS:
Occasional weak parhelia at most, rarely any other halo effects.

SEE ALSO:
anticyclonic weather (p.212); depressions (p.214); parhelion (p.138)

Although all cirrostratus has a tendency to exhibit a certain fibrous structure, on occasions this becomes extremely marked, sufficient to warrant the cloud being called cirrostratus fibratus. In general, by this stage the cloud has usually become too thick to produce any halo phenomena, although there may be occasional glimpses of a parhelion.

Cirrostratus fibratus is frequently one stage in the growth of cirriform clouds at a warm front, but it may also occur under quiet conditions when neither cyclonic or anticyclonic weather predominates.

Cirrostratus undulatus

Abbreviation: Cs un

ID FACT FILE

APPEARANCE: Relatively shallow undulations in a sheet of cirrostratus.

OCCURRENCE: With wind shear at the relevant level.

OPTICAL EFFECTS: No specific effects, but the cloud sheet may occasionally show weak parhelia.

SEE ALSO: *billows (p.100); parhelion (p.138); wave clouds (p.102)*

Occasionally a sheet of cirrostratus develops a pattern of shallow waves (billows) and may then be termed cirrostratus undulatus. The billows are often difficult to see during the day, but become prominent when illuminated by grazing light after sunset or shortly before sunrise, when they may give rise to exceptionally beautiful skies.

Such undulations may be wave clouds caused by distant mountains, but such waves generally die away downwind. In most cases, however, the undulations – like most billows – arise from wind shear at cloud level, and are approximately perpendicular to the principal wind direction.

Cumulonimbus

Abbreviation: Cb

ID FACT FILE

Appearance: Massive clouds, brilliantly white in sunshine, very dark where shadowed.

Occurrence: When active convection has raised the top of the cloud to the freezing level.

Species: calvus (p.76); capillatus (p.77)

Varieties: None

Accessory clouds: pannus (p.78); pileus (p.79); velum (p.80)

Supplementary features: arcus (p.81); incus (p.82); mamma (p.84); praecipitatio (p.85); tuba (p.86); virga (p.88)

See also: *cumulus congestus (p.32); mamma (p.84); rain (p.172); virga (p.88)*

Cumulonimbus clouds are an extremely important type, because they may produce very heavy showers of rain or snow, hail, thunder and lightning, violent winds, and (in the most intense systems) tornadoes. They are also associated with the ultimate storms: tropical cyclones.

In appearance, cumulonimbus are massive clouds, normally towering high into the sky and covering an area of perhaps 5 kilometres across. They consist of clusters of innumerable individual thermals, and have dense, 'cauliflower' heads that are brilliantly white where illuminated by sunshine. In contrast, because the clouds are so dense, little light penetrates into them, and other regions may appear extremely dark – either a dark grey or even black. Their bases appear ragged because of their precipitation in the form of heavy rain, hail, or snow, which is often seen as dense virga (fallstreaks).

Cumulonimbus may be an extremely deep cloud, especially in summer, when the surface of the ground is strongly heated by the Sun, producing clusters of thermals that grow swiftly into cumulus congestus and then transform into cumulonimbus. They often reach through all three height zones, and generally only cease to rise when they encounter the tropopause, where they spread out sideways, and often produce the

characteristic anvil shape (cumulonimbus incus). Frequently the convection is so vigorous that it actually overshoots the base of the inversion at the tropopause (in what is known as an overshooting head) before falling back to the level of the main cloud-top.

In winter, cumulonimbus clouds may be relatively shallow. This is sometimes because the tropopause is low – particularly at fairly high latitudes – but is more frequently because, with the lower surface temperatures, the freezing level is much lower in the atmosphere. It is this freezing level that is all-important in the development of cumulonimbus clouds. The cloud changes from being a cumulus congestus into a cumulonimbus when glaciation (freezing) occurs in the highest regions. This cloud then starts to produce precipitation, either in the form of rain or snow. In particularly vigorous clouds hail may also be produced.

Cumulonimbus has two species, calvus and capillatus, and these are related to the glaciation process. There are no varieties, but all of the accessory clouds and supplementary features are found with cumulonimbus.

Cumulonimbus calvus

Abbreviation: Cb cal

ID FACT FILE

APPEARANCE:
Highest cloud towers lose their hard appearance, becoming softer and smoother.

The active towers of cumulus congestus clouds have a well-defined, hard outline. When glaciation begins these towers lose their hard appearance and become slightly softer and smoother. The cloud has made the transition to cumulonimbus calvus ('calvus' is the Latin for 'bald'). This change is sometimes difficult to see unless the cloud-tops are examined with binoculars. At the same time, precipitation begins to fall from the cloud and this may be large raindrops and even hail.

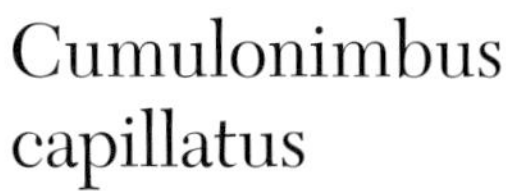

Cumulonimbus capillatus

Abbreviation: Cb cap

ID FACT FILE

APPEARANCE: High cloud towers become distinctly fibrous or striated in appearance.

When glaciation is well-advanced, the tops of the cloud towers, now consisting of large numbers of ice crystals, become fibrous or striated in appearance. The cloud has now become a cumulonimbus capillatus ('capillatus' being the Latin for 'hairy'), and may go on to develop well-defined virga or mamma. The ice crystals may be drawn out into a cirrus plume or form a distinct anvil. The precipitation is likely to be heavy rain or hail.

Pannus

Abbreviation: pan

ID FACT FILE

APPEARANCE: Fragments of ragged cloud, generally dark because shadowed by higher cloud.

OCCURRENCE: Beneath precipitating, cumulus congestus, cumulonimbus, altostratus or nimbostratus.

SEE ALSO: *altostratus (p.52); cumulus fractus (p.26); nimbostratus (p.56); stratus fractus (p.36)*

Pannus consists of ragged fragments of cloud that lie beneath cumulus, cumulonimbus, altostratus, or nimbostratus. It forms when precipitation falls from the overlying cloud into clear air beneath, where it evaporates. This cools the lower layer and increases its humidity, so that if turbulence lifts parcels of that air, it may reach its condensation point. Pannus may, therefore, be regarded as akin to cumulus fractus or stratus fractus. If the process of cooling continues for long enough, the pannus may become very extensive, and tend to merge with the cloud above.

Pileus – cap cloud

Abbreviation: pil

ID FACT FILE

APPEARANCE:
A smooth hood or cap of cloud above a rising cumuliform cloud tower.

OCCURRENCE:
When a humid layer (or layers) is lifted above its dewpoint by a rising cloud tower.

SEE ALSO:
altocumulus lenticularis (p.50); cumulus congestus (p.32); cumulonimbus (p.74)

Pileus is a distinctive form of cloud that forms above a vigorously growing cumulus or cumulonimbus cloud. It occurs when the growing cloud tower encounters a stable layer of humid air, which it lifts above its condensation level. This produces a hood or cap of cloud above the rising tower. (On rare occasions, multiple layers may be seen.) Although similar to a lenticular cloud, pileus are much shorter-lived. The rising air generally penetrates through the humid layer, which may briefly appear as a collar around the tower. The mixing within the growing tower and the invisible circulation outside it quickly cause the pileus to merge with the main cloud.

Velum

Abbreviation: vel

ID FACT FILE

APPEARANCE:
A smooth layer or layers of cloud, which generally trail behind a cumulus congestus or cumulonimbus.

OCCURRENCE:
A stable humid layer (or layers) that persists after uplift.

SEE ALSO:
cumulus congestus (p.32); cumulonimbus (p.74); pileus (p.79)

Active cumulus congestus and cumulonimbus clouds often cause uplift over a considerable area, leading to a broad sheet of thin stratiform cloud becoming visible, in a manner similar to that found with pileus. This layer is sometimes draped over the rising heads but more frequently they penetrate the cloud sheet and continue rising into the clear air above it. The sheet of cloud is known as velum (the Latin for 'sail' or 'veil'), and tends to trail behind the active cells. It may be extremely persistent, lasting long after the cumulonimbus has passed into the distance. Quite frequently, more than one layer of cloud is created in this way.

Arcus

Abbreviation: arc

ID FACT FILE

APPEARANCE:
A dark arch, wedge, or roll of cloud ahead of an advancing cumulonimbus, multicell or supercell system.

OCCURRENCE:
When cold air flowing out of the active system undercuts warmer surrounding air.

SEE ALSO:
cumulonimbus (p.74); multicell storms (p.227); storms (p.226); supercell storms (p.227)

The supplementary feature known as arcus is a dense line or roll of cloud that sometimes forms the low, leading edge of an active cumulonimbus cloud or one of the larger multicell or supercell systems. It may also be seen occasionally with less vigorous cumulus clouds. The cloud may appear as a dark arch across the sky, as a distinct wedge-shaped mass, undercutting the surrounding air (a 'shelf cloud'), or as a cloud mass detached from the main cloud, that rotates around a horizontal axis (a 'roll cloud').

Incus

Abbreviation: inc

ID FACT FILE

APPEARANCE:
A flattened anvil top to cumulonimbus clouds.

OCCURRENCE:
When vigorous convection reaches an inversion.

SEE ALSO:
cirrus (p.58); cumulonimbus (p.74); multicell storms (p.227); supercell storms (p.227)

When cumulonimbus clouds reach an inversion (often the tropopause) their upwards growth is checked. The great quantities of ice crystals formed in the upper regions of the cloud are spread out by the winds at that altitude (which are generally stronger than those at lower levels). The result may be a vast cirrus plume stretching far downwind or an anvil shape (incus). Although most of the ice crystals are carried downwind, the outflow is often strong enough for some of the air to extend upwind, producing the highly characteristic anvil profile. The undersides of the overhanging shelf of cloud often exhibit virga and mamma.

Anvils sometimes look smooth, but the ice crystals generally give them a striated

appearance. Depending on the exact conditions they may be extremely persistent and may last after the parent cumulonimbus has decayed. Such detached anvils basically consist of cirrus spissatus and appear very dark against the light. Remnants of anvils and cirrus plumes may persist for days as patches of disorganized cirrus.

When a cumulonimbus incus is at a considerable distance, where the anvil itself does not block the view, it is sometimes possible to see that some of the most vigorous updraughts have overshot the inversion, producing a dome of cloud (an overshooting head) at a higher level. The air in such heads subsides back to the level of the inversion and spreads out from the cloud in all directions.

Mamma

Abbreviation: mam

ID FACT FILE

APPEARANCE: Rounded masses beneath higher cloud.

OCCURRENCE: Beneath cumulonimbus incus, and also altocumulus, altostratus, cirrus, cirrocumulus, and stratocumulus.

SEE ALSO: *cumulonimbus (p.74); virga (p.88)*

Mamma are rounded masses of cloud that hang beneath other clouds. They are most prominent underneath cumulonimbus anvils, but also occur with altocumulus, altostratus, cirrus, cirrocumulus, and stratocumulus. They occur when cool, humid air descends into a warmer layer and causes the latter to cool to its dewpoint and condense into droplets. Warm air tends to rise between the bulges to compensate for the downdraughts. Sometimes the bulges are smooth, globular masses, but quite often the mamma assume the shape of long, contorted tubes. In cirrus they give the streamers a scalloped appearance.

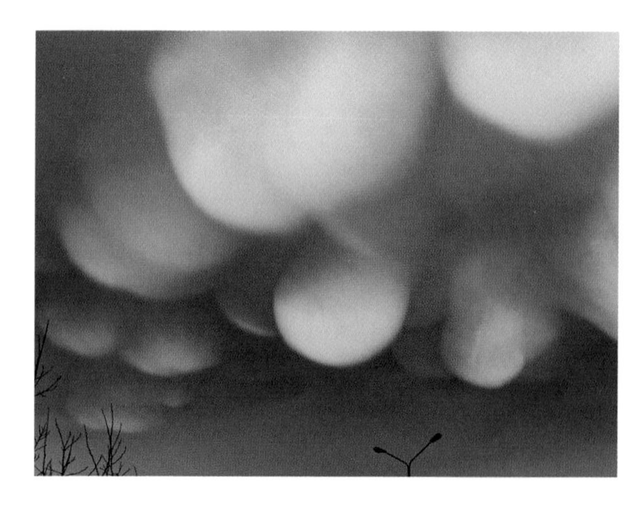

Praecipitatio

Abbreviation: pra

ID FACT FILE

Appearance:
Streaks of precipitation reaching the ground, generally grey for rain, and brighter for snow and hail.

Occurrence:
Below cumulus congestus, cumulonimbus, and nimbostratus in particular, but also altostratus, stratocumulus, and stratus.

See also:
hail (p.178); rain (p.172); snow (p.176); virga (p.88)

The term praecipitatio is used for a cloud type that is producing any form of precipitation (whether solid or liquid) that actually reaches the ground. (The term virga is used if the precipitation evaporates in mid-air.) Praecipitatio therefore covers everything from a fall of tiny ice crystals or a light dusting of snow to the largest hail, and from an almost imperceptible drizzle to a torrential downpour. Although it obviously applies to cumulus congestus, cumulonimbus, and nimbostratus, under certain circumstances it may also be used for altostratus, stratocumulus, and stratus, even though, in all three cases, the amount reaching the surface may be very small.

Tuba - funnel cloud

Abbreviation: tub

ID FACT FILE

APPEARANCE:
A narrow tubular or conical cloud that hangs down from more massive higher cloud, but does not reach the surface. The colour may be anything between white and dark grey, depending on the lighting.

OCCURRENCE:
Beneath cumulus congestus or cumulonimbus clouds, on gust fronts, and associated with severe storm systems, including hurricanes.

SEE ALSO:
cumulus congestus (p.32); cumulonimbus (p.74); thunderstorms (p.226); tornadoes (p.234); waterspouts (p.232)

The term 'tuba' is the formal meteorological name for a funnel cloud that hangs down from the clouds, but does not touch the surface. Among the general public such forms are often described as 'tornadoes' or 'waterspouts', but these have more specific meanings, which are described later. Tubas are actually quite common, especially as they are generated by many vigorous storm systems, as well as under certain specific, but calmer, conditions.

Tubas usually start as small projections from the base of larger clouds, in particular from cumulus congestus or cumulonimbus, where they are created by strong downdraughts. They may then grow down towards the surface. (If they do touch the ground, then they become true landspouts, waterspouts, or tornadoes, but the last of these form by a completely different process.) The lifetime of such funnel clouds is usually just a few minutes.

Virga

Abbreviation: vir

ID FACT FILE

APPEARANCE:
Trails of precipitation beneath a cloud or other source. Usually, but not always, curved, and do not reach the ground.

OCCURRENCE:
Beneath altocumulus, altostratus, cumulus congestus, cumulonimbus, or nimbostratus clouds; below contrails, distrails, and fallstreak holes.

SEE ALSO:
condensation trails (p.100); dissipation trails (p.110); cumulus congestus (p.32); cumulonimbus (p.74); fallstreak holes (p.106); praecipitatio (p.85)

Many clouds show streaks of precipitation beneath them, and if these streaks do not reach the ground they are known as virga. (They are also frequently called fallstreaks.) They may consist either of water droplets (raindrops) or ice crystals – normally in the form of snowflakes. In many cases, of course, ice crystals fall from the cloud, but then melt to water droplets, which, in turn, evaporate. As droplets evaporate, they naturally become smaller and fall more slowly, trailing behind the cloud. In contrast, compact ice crystals (rather than snowflakes) fall almost vertically, but there is often a distinct hook to the virga where they are melting and starting to evaporate and, as a result, are beginning to shrink in size.

Although the virga that accompany cumulus congestus, cumulonimbus, and nimbostratus clouds are probably the most frequent, they

are, at times, associated with many other clouds. They are commonly seen with altocumulus castellanus and altocumulus floccus, while cirrus, for example, could be regarded as nothing but virga, falling from the generating heads that are sometimes clearly visible. The striations that appear at the top of a cumulonimbus cloud as it changes into cumulonimbus capillatus are actually a sign of falling ice crystals, and are thus a form of virga.

Virga are also frequently seen below aircraft condensation trails, and are normally visible below the approximately circular holes that sometimes appear in cirrocumulus and altocumulus, known as fallstreak holes, as well as below the less common dissipation trails.

Cloud formation

Clouds, mist and fog all form when air (which always contains some moisture) cools below its dewpoint, giving rise to tiny cloud droplets. This cooling may occur through two main mechanisms: either the air passes over a cold surface, or it rises and cools.

The first process is responsible for certain forms of mist and fog (known as advection fog, p.184). This occurs when humid air is carried gently across cold ground, particularly when the ground is covered in snow or ice. If the wind is slightly higher, turbulence mixes the cooled air into the lowest layer of the atmosphere, giving rise to a low blanket of stratus cloud. On rare occasions, the situation may be reversed, with very cold, dry air passing over relatively warm, unfrozen water. Water molecules evaporate into the dry air, where they are almost instantly cooled and condense, producing steam fog (otherwise known as arctic sea smoke).

There are various mechanisms that cause air to rise, but there are three particularly important ones:

- by convection, normally through heating of the lowermost layers by warmth from the surface, particularly from the land

- by forced ascent over hills or mountains (known as orographic lifting)

- by lifting at a frontal surface, where warm air is being undercut by cooler air

The last of these three mechanisms primarily occurs in depressions, but a similar form of lifting takes place at line squalls and in association with the cold outflows from cumulonimbus clouds and storm systems.

In addition, a process known as convergence, where air is forced into a restricted volume and rises to escape, in also involved. Convergence occurs, for example, in the centre of depressions, where the air rises and flows outwards at higher levels.

Convective clouds

When the ground is heated by the Sun, bubbles of air (known as thermals) become warm enough to break away from the surface and start to rise. There is a circulation in these individual thermals, with warm air rising in the centre and cooler air descending around the outside. Early in the day, the thermals mix with the surrounding air and die away, but later when temperatures are higher they rise higher, the air expands (because the pressure decreases with height), and eventually cools sufficiently for its water vapour to condense into cloud particles. This occurs at what is known as the condensation level.

All cumuliform clouds, from the smallest tufts of cumulus fractus to the largest cumulonimbus clouds, basically arise in this

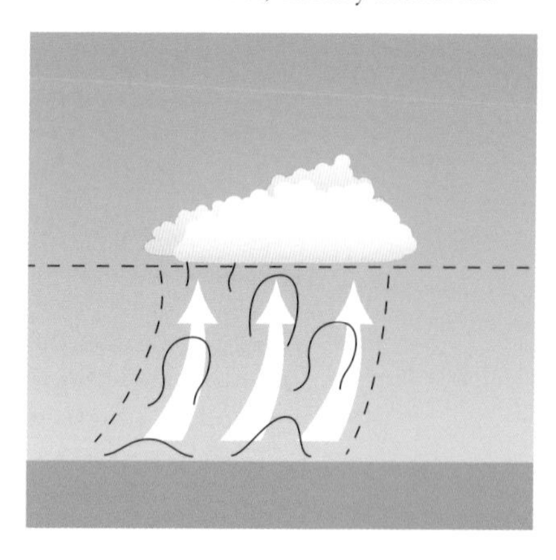

manner, so the process is an extremely important one for a wide range of atmospheric events. In general, cumuliform clouds are an indication of instability (p.13) – at least in the lower levels of the atmosphere. If the rising air encounters an inversion (a region of stability) cumulus may spread out sideways into a layer of stratocumulus or altocumulus, unless convection is strong enough to force a way through the stable layer and continue upward growth.

Orographic clouds

If the wind encounters a barrier, such as a mountain range, the air will again be forced to rise, expand, and cool, and may then reach its dewpoint at the condensation level. In this case, however, the type of cloud that forms will depend upon the exact conditions that prevail. If the air is stable, it will tend to sink back to its original level after passing the obstacle. The cloud will be stratiform in nature: stratus, nimbostratus, altostratus, or even cirrostratus. It will lie only over the higher ground and not extend far downwind, unless the wind and obstacle combine to create a series of wave clouds (described later).

If the air is unstable (or becomes so when forced to rise), the clouds will be cumuliform in nature, in particular, cumulus and cumulonimbus. Normally the wind carries

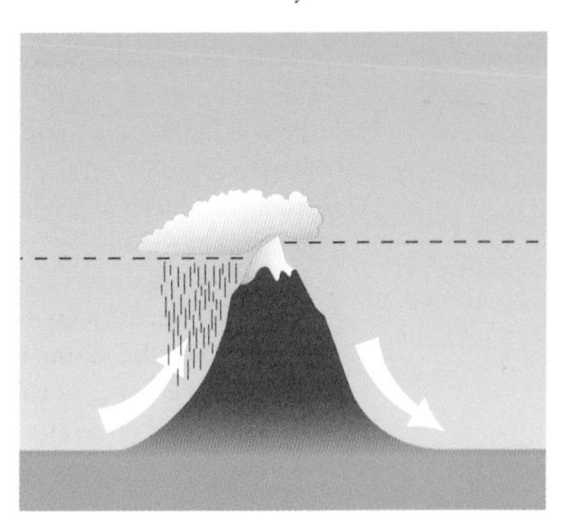

these downwind, to be replaced by new cloud, but occasionally conditions may cause the cloud to remain over the high ground, where cumulonimbus may produce prolonged heavy rainfall.

In general, rainfall induced by the high ground is heaviest on the windward slopes and over the peaks, declining to leeward, where the air descends and warms, giving rise to a 'rain shadow' with lesser rainfall on the downwind side.

Frontal cloud

Fronts arise where two air masses with different characteristics come into contact, most particularly in depressions. In general, the colder air mass undercuts the warmer air, lifting it away from the surface, where it cools and produces cloud. The type of cloud formed depends on the type of front. At a warm front *(above)*, the warm air is advancing and slides above the cold, giving rise to an inversion and stable conditions, the clouds are stratiform in nature: cirrostratus, altostratus, and nimbostratus.

At a cold front, the cold air is advancing and undercuts the warm air. Sometimes the clouds produced are very similar to those at a warm front (particularly in winter at middle latitudes), but because cold air is generally advancing over a warm surface, unstable

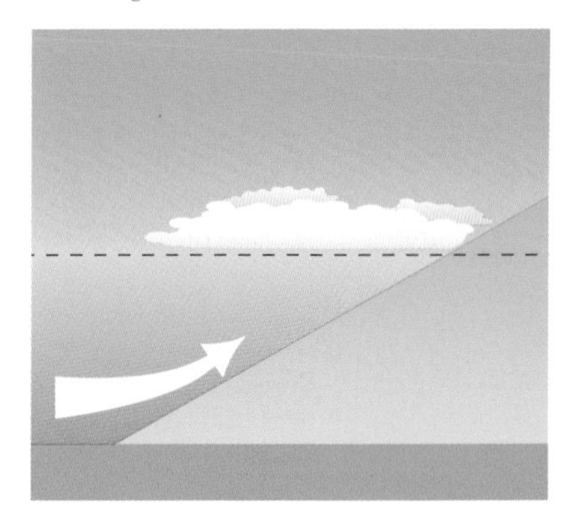

conditions result, producing vigorous convection in the form of cumulonimbus clouds and heavy rainfall.

Conditions are rarely clear-cut, however, and many fronts show a mixture of cumuliform and stratiform clouds. Convective clouds are sometimes embedded in warm fronts, but it is much more common to find extensive stratiform clouds associated with cold or occluded fronts. This is particularly the case when the air at middle levels in a depression has a tendency to sink – a situation that is quite common in maritime regions. This suppresses convection at the cold front and may produce thick stratocumulus rather than the cumuliform clouds one normally associates with cold fronts. (A similar process may occur at warm fronts and suppress the normal sequence of cirrostratus, altostratus and nimbostratus, producing thick stratocumulus and stratus instead.)

Cloud streets

ID FACT FILE

APPEARANCE:
Long lines of cumulus cloud extending downwind. The clear lanes are generally at least twice the width of a line of cloud.

OCCURRENCE:
Generally within a restricted layer, and where the sources of thermals persist for some time.

SEE ALSO:
cumulus (p.24); convection (p.92)

Once convection has become established, it is quite common for lines of cumulus to arise. These cloud streets appear to converge in the distance because of perspective, and are therefore described as cumulus radiatus. They occur when specific sources of warmth persist to feed new thermals. When there is just one source of heat (such as a mountain peak or escarpment heated by the Sun), a single, long cloud street may be produced. This is frequently seen downwind of isolated oceanic islands.

More commonly, however, there are several parallel cloud streets, which form most readily when the convective layer is relatively shallow, topped by an inversion, as often happens when there are anticyclonic conditions with air

subsiding from above, or when radiation fog has formed overnight. Air rises as far as the inversion along the lines of cloud and subsides in the clear spaces between them. The spacing is about two to three times the depth of the convective layer.

Cloud streets may persist all day if the individual cumulus are short-lived, but if the air is humid, the cumulus may be slow to evaporate and spread out into stratocumulus. Similarly, convection needs to be relatively weak. If it is strong enough to break through the inversion, the regular pattern of ascending and descending air will be disrupted and the cloud streets will disappear.

Billows

ID FACT FILE

APPEARANCE:
Regularly spaced rolls or streaks of clouds, generally perpendicular to the wind direction.

OCCURRENCE:
When wind shear occurs in adjacent layers of air or when a stable layer is forced to rise.

SEE ALSO:
fog (p.184); wave clouds (p.102)

Billows (technically known as the undulatus variety) are a common type of cloud structure, where a layer is broken up into regularly spaced rolls of cloud. They generally arise when there is wind shear between two adjacent layers in the atmosphere, that is, when there is a difference in either the wind speed or in the direction (or both) in the two layers. The billows run approximately at right-angles to the wind. Sometimes the billows are not separated by clear air, and then appear from above as an undulating surface, and from below as thicker and thinner patches of cloud.

When a stable layer is forced to rise it may create billows that resemble breaking waves, although the mechanism is actually completely different. Such billows are extremely short-lived, lasting no more than a few minutes.

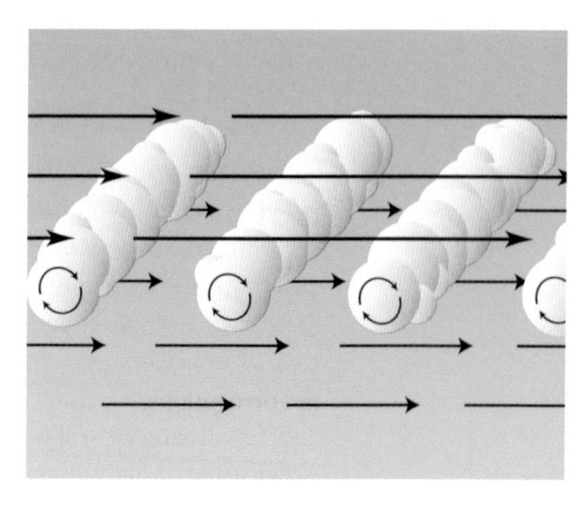

Although billows often bear a striking resemblance to wave clouds, there is one all-important difference: billows move with the cloud layer, whereas wave clouds remain stationary in the sky. Wave clouds may themselves sometimes have a billow-type structure. Allied to billows, and sometimes appearing with them, are finer undulations, parallel to the wind direction, which are known as corrugations.

Wave clouds

ID FACT FILE

APPEARANCE:
Smooth, lens- or almond-shaped clouds that remain stationary in the sky.

OCCURRENCE:
When a steady wind encounters a barrier, setting up a train of waves. Clouds may be created at considerable distances, where the high ground itself may be out of sight.

SEE ALSO:
altocumulus (p.46), cirrocumulus (p.66), instability (p.12), nacreous clouds (p.116), noctilucent clouds (p.118), stability (p.12), stratiform clouds (p.15), stratocumulus (p.40)

When the wind encounters an obstacle, it sets up a train of waves in the atmosphere. Sometimes these waves are invisible, but frequently they are sufficiently large to raise a humid layer of air above its dewpoint, causing clouds to form at the wave-crests. In some cases, a whole series of clouds may be produced extending far downwind.

Depending on conditions, such wave clouds (also known as lenticular clouds) may occur as stratocumulus lenticularis, altocumulus lenticularis, or cirrocumulus lenticularis. Wave clouds occur under stable conditions, when air, forced to rise over a barrier, automatically tends to return to its previous level. Despite being called '-cumulus', lenticular clouds are actually stratiform (layer) clouds of a very specific type.

Wave clouds always appear smooth and almond- or lens-shaped. They remain stationary in the sky and air flows through them, cloud particles condensing at the leading

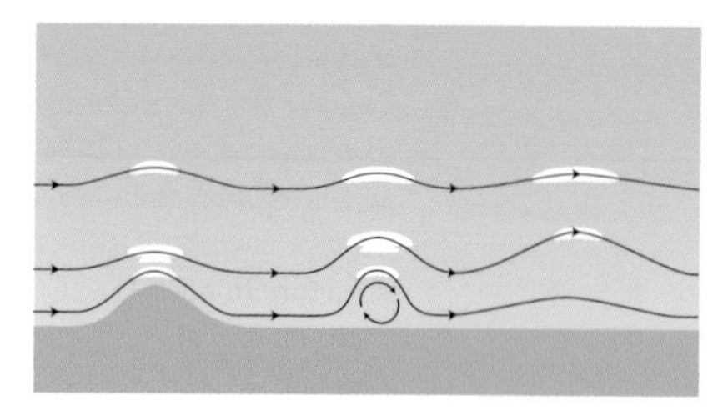

edge, and evaporating at the trailing edge. This process may often be seen if the clouds are examined with binoculars. A train of individual clouds may form over an isolated peak, but with a line of mountains, wave clouds often merge to produce lines of cloud parallel to the range. Wave clouds are persistent and may last for hours, provided the wind direction and strength remain unchanged.

Frequently there is more than one humid layer of air, so a series of clouds may form above one another. Such stacks of clouds are known as 'piles d'assiettes' ('piles of plates') and may appear extremely striking, especially when illuminated by a low Sun.

Sometimes the amplitude of the waves is extremely large and they extend to great heights. (Such conditions are valued by glider pilots seeking to achieve maximum altitude.) Nacreous clouds are wave clouds, but the effects may extend even farther. Much of the structure in noctilucent clouds arises from wave motion at their heights of around 80 kilometres.

Pyrocumulus

Clouds frequently form over the rising columns of air from fires, both natural and man-made ones. Such clouds are known by the rather clumsy name of 'pyrocumulus', from the Greek word for 'fire' and the Latin word for 'heap'. Even if the column of smoke cannot be seen clearly, the base of such clouds is often a darker, or brown shade, that tends to distinguish them from ordinary cumulus.

These clouds will form over surprisingly small fires – under suitable conditions they have even been seen over large garden bonfires. The majority are like ordinary cumulus humilis or cumulus mediocris. Over large wild-fires, however, such as the widespread fires in Florida some years ago *(opposite)*, they may give rise to cumulus congestus or even cumulonimbus. These last two types of cloud

ID FACT FILE

APPEARANCE:
Cumulus cloud, usually with a base that is darker than normal, or brown in colour. The associated smoke column is not always visible.

OCCURRENCE:
Over any form of fire, and usually when the condensation level is fairly low. Strong winds, with the resulting turbulence, sometimes prevent their formation.

SEE ALSO:
cloud formation (p.90); cumulus (p.24); cumulonimbus (p.74); lightning (p.228)

may, of course, produce rain, and it has been known for rainfall to extinguish the fire that created the cloud. In most cases, however, the clouds have drifted downwind, and precipitation falls elsewhere. In some extreme cases, the cumulonimbus may produce thunderstorms, and on a few occasions lighting from one fire-generated cumulonimbus has started wild-fires in a different location.

The strong updraughts produced by wildfires sometimes create intense tornado-like fire devils (p.230) and on occasions these have been observed to be capped with their own individual, pyrocumulus clouds.

Fallstreak holes

ID FACT FILE

APPEARANCE:
Large holes that suddenly appear in a layer of altocumulus, cirrostratus or (occasionally) stratocumulus.

OCCURRENCE:
Formation appears to be random, although a series of holes may be related to slight undulations in the parent layer.

SEE ALSO:
distrails (p.108); supercooling (p.172); virga (p.88); wave clouds (p.102)

Layers of altocumulus and cirrocumulus sometimes develop large holes. These fallstreak holes are much larger, more symmetrical, and differ greatly in appearance from the gaps that are found in altocumulus or cirrocumulus lacunosus. Very frequently, a trail of cirrus is seen beneath a hole, and this gives a clue as to its formation.

The water droplets in these types of cloud are supercooled, and if they freeze (i.e., if the cloud becomes glaciated), the ice crystals fall out, creating the hole. The trigger may be ice crystals falling from a yet higher layer of cloud. The reasons for the striking symmetry of many fallstreak holes are uncertain. It is thought that the ice crystals break into tiny fragments, which then act as freezing nuclei for further crystals, which then themselves fragment, in a

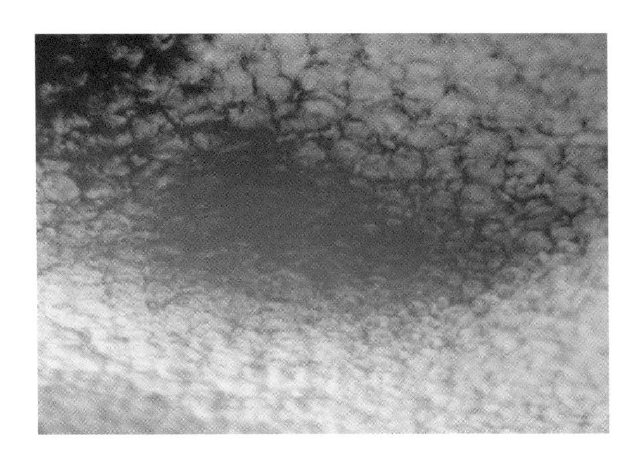

runaway process, rapidly producing the hole and the clumps of virga hanging beneath it. Very occasionally, a remarkably regular series of holes may appear in a cloud, and these are possibly related to slight undulations, like those that create wave clouds.

A similar process is thought to be responsible for most of the dissipation trails (distrails) that aircraft produce in thin cloud.

Fallstreak holes should not be confused with the gaps where altocumulus or cirrocumulus have not formed, or are gradually being eroded by evaporation of the cloud droplets.

Contrails and distrails

ID FACT FILE

SEE ALSO:
mamma (p.84); supercooling (p.172); virga (p.88)

Condensation trails (contrails) are able to provide information about the way in which the weather may develop. They are, of course, lines of cloud formed by the condensation or freezing of the water vapour produced by the aircraft engines. Because condensation or freezing is not instantaneous, there is always a gap between the engines and the start of the contrail.

Seen from underneath, contrails consist of twin lines of cloud. The exhaust gases are drawn into the centres of invisible vortices shed by the two wingtips. So there are always twin trails, regardless of the number of engines. Seen from the side, contrails show a series of bulges, rather like the mamma seen beneath

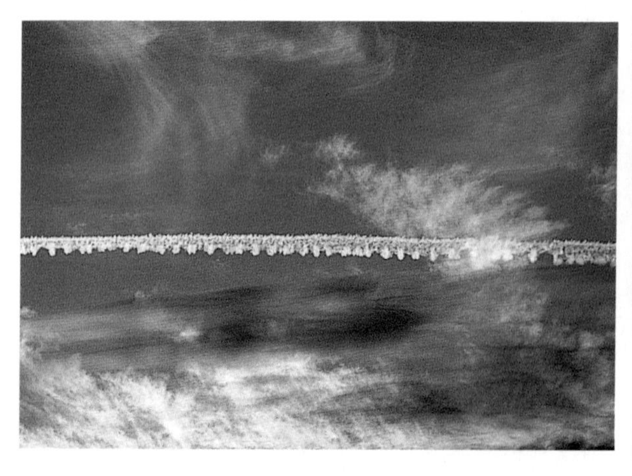

certain types of cloud. The exhaust is hot, and thus tends to rise, but the air is forced down by the passage of the aircraft, and the end result is the bulges that are often seen.

Contrails' persistence depends upon the water-vapour content at the aircraft's height. When the air is dry, as it often is at high altitudes, contrails evaporate or (if they consist of ice crystals) sublime, and disappear rapidly. When the air is humid, such as in advance of a warm front, they may spread and remain extremely persistent. In some cases they may expand greatly to cover the whole sky with a blanket of cirriform cloud.

Just as aircraft can produce cloud, they can also disperse it. Such dissipation trails (distrails) do not occur as frequently as contrails, but are still fairly common. They are produced when an aircraft flies at the same

altitude (or slightly above) a thin layer of cloud. There are at least three factors that can cause the cloud to dissipate: the aircraft may mix dry air from a slightly higher layer into the cloud; the particles emitted in the exhaust may act as freezing nuclei for supercooled water droplets; and the heat from the exhaust may be sufficient to cause cloud droplets to evaporate. The first two mechanisms are the most likely, and the second (freezing of existing supercooled droplets) causes ice crystals to fall out of the cloud. Trails of virga are often seen beneath distrails.

Banner cloud

ID FACT FILE

APPEARANCE:
A plume of cloud hanging behind a mountain peak, often unaccompanied by other clouds.

OCCURRENCE:
With a steady (and fairly strong) wind from a constant direction from the opposite side of the mountain that is warmed by the Sun.

SEE ALSO:
cloud formation (p.90); pressure (p.206); valley winds (p.198)

A plume of cloud sometimes hangs downwind of a steep mountain peak. These banner clouds are particularly well-known behind the Matterhorn on the Swiss-Italian border *(below)*, and Mount Everest, but often occur elsewhere, where there are isolated peaks, such as the Rock of Gibraltar. Generally only one peak in a mountain range has a banner cloud, but on rare occasions more than one have been seen at the same time.

Banner clouds require a complex set of circumstances to form. Generally, one steep face needs to be heated by the Sun, and a fairly strong wind must be blowing from the opposite (and therefore shaded and cooler) side of the mountain. The wind creates a low-pressure region behind the peak. It also produces

eddies, which combine with the heating effect to draw warm air up the sunward face. The rising air expands and cools (helped by the slightly lower pressure behind the peak), reaches its dewpoint, and the moisture condenses to produce the cloud.

Banner clouds may persist for many hours, even though they may cast a shadow and reduce the heating on the sunward side of the peak. Generally they last throughout the day, providing there is no major shift in wind direction. Like many other clouds, however, they die away when warming ceases with nightfall.

Jet-stream clouds

ID FACT FILE

APPEARANCE:
High-level cirrus, generally in elongated parallel bands, frequently with billows perpendicular to the general flow. The motion along the jet is normally readily apparent.

OCCURRENCE:
Often appears ahead of an approaching depression, and may sometimes be observed after a depression has passed.

SEE ALSO:
billows (p.100); cirrus (p.58); depressions (p.214); jet streams (p.189)

The westerly jet streams that flow around the Earth have a powerful influence on the development of depressions and the weather. They are normally invisible, but sometimes highly distinctive clouds occur within them. At the altitudes where the strongest jet streams occur, just below the tropopause, temperatures are very low, so the clouds consist of cirrus. Even though they are at high altitude, the wind speeds are so high, 100 knots (about 185 kph) or even more, that the movement of the clouds themselves is readily visible.

The bands of jet-stream cirrus generally appear as streaks elongated along the jet, so they may be classified as cirrus radiatus.
Billows are often seen perpendicular to the airflow, and these may be regarded as cirrus undulatus.

In the northern hemisphere, the jet stream roughly parallels the polar front and lies in a flattened 'S-shape' above depressions (a 'Z' in the southern hemisphere). Very often the first sign of an approaching warm front is jet-stream cirrus racing across the sky towards the east or southeast.

Billows are an indication of wind shear (p.100), which may be extreme on the edges of jet streams. (The maximum wind speed recorded in a jet stream is 656 kph (354 knots) over the Outer Hebrides, 13 December 1967.) When no clouds are present, these are the regions where aircraft encounter clear-air turbulence.

Nacreous clouds

ID FACT FILE

APPEARANCE:
Clouds exhibiting brilliant spectral colours, seen after sunset or before sunrise. On rare occasions, luminous white forms are seen.

OCCURRENCE:
Most often seen at latitudes greater than 50°N or S. Sometimes seen very briefly from lower latitudes.

SEE ALSO:
iridescence (p.130); noctilucent clouds (p.118)

Occasionally, shortly after sunset or before sunrise, extremely striking, brilliantly coloured clouds appear high in the atmosphere, shining like mother-of-pearl (whence their name). These clouds are actually in the stratosphere, at heights of 15–30 km, where they are illuminated by the Sun, while observers on the ground are in darkness. The beautiful colours of evening displays frequently shift towards orange and red as the Sun slips farther below the horizon, and a reversed sequence occurs at dawn. The colours (caused by iridescence) set them apart from noctilucent clouds, which are also most prominent around midnight.

Nacreous clouds are most readily seen at latitudes greater than about 50°, although they may sometimes be glimpsed fleetingly from lower latitudes. The brilliant colours, like those

seen in iridescence, arise from diffraction by tiny, evenly sized cloud particles. Close to the poles, larger particles may form with the slowly falling temperatures at the onset of winter, and these produce rarer displays of brilliant white clouds, which seem to be more common over Antarctica than in the north.

These two types of nacreous cloud, and another unrelated type (not visible from the ground), are also known as polar stratospheric clouds. All consist of ice crystals with sulphuric-acid droplets as initial freezing nuclei. The chemical reactions that destroy ozone and create the ozone holes take place on the surfaces of the cloud particles.

Noctilucent clouds

ID FACT FILE

APPEARANCE:
Silvery-white or yellowish cirrus-like clouds visible around midnight towards the pole.

OCCURRENCE:
Most often seen at latitudes greater than 45°N or S. Sometimes glimpsed from lower latitudes.

SEE ALSO:
cirrus (p.58); nacreous clouds (p.116); wave clouds (p.102)

During the period of about two months around midsummer, silvery-white or slightly yellowish clouds may be seen during the night in the general direction of the pole, by observers at moderately high latitudes (above 45°N and S). These noctilucent clouds lie at heights of about 80 km, far higher than any other clouds, and they remain in sunlight throughout the night, when even nacreous clouds (at 15–30 km) are in darkness.

Noctilucent clouds could be mistaken for cirrus. They often show bands, billows and other structures, which frequently move westwards during the night. In fact, they

consist of a very thin layer and most of the apparent structure is caused by undulations in the cloud sheet, which mean that more particles lie along certain lines of sight. Rather like wave clouds, the undulations appear to be related to mountain ranges on the ground far beneath them.

The precise mechanisms by which noctilucent clouds are formed are still a subject of debate, but the clouds themselves seem to be becoming more frequent. It is not known whether this is perhaps caused by global warming, or by an increase in the numbers of informed observers.

Rocket and meteor trails

ID FACT FILE

APPEARANCE:
Generally narrow trails, sometimes highly coloured, becoming distorted with time.

OCCURRENCE:
In upper atmosphere, at random times when caused by meteoroids, but at very specific times when produced by rocket launches or satellite re-entries.

There are various events that may occur in the upper atmosphere, and which may sometimes cause confusion. In particular, rocket exhaust trails, and the trails left by certain meteors – particularly the extremely bright ones known as fireballs – may easily be mistaken for nacreous or noctilucent clouds.

Rocket trails normally appear as narrow streaks, and these often display iridescence, similar to nacreous clouds. Occasionally this may be extremely spectacular, as may the colours exhibited by chemicals that are sometimes released in upper-atmosphere experiments.

SEE ALSO:
iridescence (p.130); nacreous clouds (p.116); noctilucent clouds (p.118)

Persistent meteor trails – known as 'trains' – may persist for many minutes after the meteor has passed. Those left by large meteors may also be very long. The distortions in such trains (and in rocket trails) provide a useful means of determining upper-atmosphere winds, which are otherwise extremely difficult to detect.

Satellite re-entries may produce similar trails. Unlike meteors, however, which are completely unpredictable, the re-entry of debris from orbit is normally known with a reasonable degree of certainty. Re-entry predictions are available over the Internet, and provide a means of determining the true cause of the event.

Aurorae

ID FACT FILE

APPEARANCE: Patches, arc, bands of light, normally appearing green or red, sometimes quiescent, but at others varying in both intensity and position.

OCCURRENCE: Generally in the part of the sky towards the pole, but sometimes seen overhead or farther towards the equator.

SEE ALSO: *noctilucent clouds (p.118)*

The aurora borealis, or Northern Lights (and its southern counterpart, the aurora australis), have always been reported by meteorological observers. Like nacreous and noctilucent clouds, they are also visible in twilight or during the night, generally appearing in the sky towards the pole.

Aurorae occur when particles from the Sun cascade into the upper atmosphere, exciting the atoms at great heights (100–1000 km and more) and causing them to emit light. The colours are highly distinctive and indicate which atoms are involved. The most common colours are green, emitted by oxygen, and red from nitrogen molecules. When the tops of displays are in sunlight, a purple-violet emission from nitrogen is sometimes seen. The visible colours are strongly dependent on the

observer's eyesight. Green displays, although seen, may appear colourless, and many people are unable to see the red hues.

There are several distinct forms:

- arc: an arch-like structure with a sharp lower edge and diffuse upper border
- band: a structure like curtains or ribbons with distinct folds, and frequently showing rapid movement
- corona: rays that appear to radiate from a point high overhead
- glow: a faint glow on the horizon towards the pole
- patch: a weakly luminous area, often looking like a faintly glowing cumulus

cloud; sometimes fluctuates in strength
- ray: a vertical streak of light often seen with arcs and bands
- veil: a very weak, even glow of light, covering a large area of sky

No two auroral displays are alike and weak displays are often taken for clouds illuminated by distant lights or (with red aurorae) for distant fires. Strong displays may begin with a quiet homogeneous arc, which develops rays, then turns into a band. A very strong event may expand towards the equator and, when it moves overhead, appear as a corona. Generally, displays break up towards dawn, degenerating into pulsating patches of light.

Optical phenomena

There is an extremely wide range of optical phenomena that may be seen in the sky, and to anyone unfamiliar with the subject there may seem to be a bewildering variety of different effects. In broad terms, however, the various effects may be sub-divided, depending on their location in the sky, and whether they show any colours or appear white. Coloration of the sky, the various forms of shadows, and abnormal refraction effects (mirages) are all dealt with separately later.

Some of these effects are given slightly different names when they are produced by light from the Moon, rather than the Sun. The lunar equivalent of a parhelion, for example, is a paraselene; moonbows are similar to rainbows; and Moon pillars are sometimes visible.

- Strongly coloured phenomena:

Rainbow	p.144
Dewbow	p.151
Iridescence	p.130
Circumhorizontal arc	p.139
Circumzenithal arc	p.139
Parhelion	p.138

- Weakly coloured phenomena:

Moonbow	p.149
Corona	p.128
Glory	p.152
Circular haloes	p.132

- Colourless (white) phenomena:

Fogbow	p.150
Heiligenschein	p.153
Parhelic circle	p.136
Various halo arcs	p.136
Sun pillar	p.142
Subsun	p.143

Although shown here as 'strongly coloured', frequently the colours of some of these phenomena may be much weaker and occasionally more-or-less absent.

Corona

ID FACT FILE

APPEARANCE: A coloured disk, ring, or rings around the Sun or Moon.

OCCURRENCE: Most commonly seen in Ac, As and Cc

SEE ALSO: *altocumulus (p.46); altostratus (p.52); iridescence (p.130)*

A corona is a coloured ring or rings surrounding the disk of the Sun or Moon as seen through thin medium or high cloud – most frequently either altostratus or altocumulus. Coronae are more often accidentally noticed around the Moon, because its light is not so dazzling, but hiding the Sun will frequently reveal a corona. Similar effects occur in mist or fog. Iridescence, with which a corona might be confused, generally lies much farther from the Sun or Moon.

Unlike haloes, which change slowly (if at all), coronae often vary in both size and shape, and also in the strength of colour as clouds move

across the Sun or Moon. The inner region, which is known as the aureole, is just a few degrees across, and shows a distinct reddish-brown outer border. It may appear bluish in the centre. Other rings may be seen farther from the centre, and these show weak spectral colour, with red on the outside. Such rings are an indication that the cloud particles are very uniform in size: the smaller the particles, the larger the rings.

Iridescence

ID FACT FILE

APPEARANCE: Coloured bands (often pastel shades) around the edges and within clouds.

OCCURRENCE: Common to many clouds, but most often seen in thin Ac, As and Cc

SEE ALSO: *altocumulus (p.46); altostratus (p.52); cirrocumulus (p.66); corona (p.128); nacreous clouds (p.116)*

Iridescence is a common optical phenomenon, but very frequently goes unnoticed, largely because it appears in clouds near the Sun (or Moon), when it is lost in the general glare. Shielding the eyes from the Sun will often reveal cloud iridescence.

It consists of bands of colour, generally running along the edges of clouds, but sometimes appearing within them. Although it is most pronounced in thin altocumulus, altostratus, and cirrocumulus, it may appear fleetingly in nearly any type of cloud as it passes across the Sun.

The colours tend to be strongest about 30–35° away from the Sun and are similar to those in a corona, although more delicate pastel shades

are also very common. The colours are produced by diffraction of the sunlight (or moonlight) by cloud particles, and strong bands of individual colours indicate where all the cloud droplets or ice particles are identical in diameter. The pastel shades are produced when there is a slight mixture of particle sizes, causing more than one colour to be present.

The most striking occurrence of iridescence occurs with the rare nacreous clouds, occasionally seen at sunset and sunrise.

Haloes

ID FACT FILE

Appearance:
Circles or arcs, centred on the Sun or Moon, often colourless or with faint red and blue tints on inner and outer edges. Often accompanied by other arcs and spots of light.

Occurrence:
Most often when a thin layer of cirrostratus covers the sky or, in polar regions, when fine ice crystals are suspended in the air.

Haloes are some of the most common optical phenomena visible in the sky, and it has been estimated that at middle latitudes they are visible about once every three days. They are relatively unknown, however, because the most common forms occur near the Sun, and are readily visible only when it is hidden.

All halo effects are produced when light is refracted through ice crystals, and at most latitudes the best conditions for their formation occur when there is a thin sheet of cirrostratus. Such clouds are so thin that many people do not realise that they are present until they become thick enough to start to reduce the heat from the Sun. If you become aware that there is a thin veil of high cloud spreading across the sky, hide the Sun with

See also:
cirrostratus (p.70); circumhorizontal arc (p.139); circumzenithal arc (p.139); parhelion (p.138)

your hand, and a halo may well appear. Such conditions frequently occur ahead of the warm front of an approaching depression.

Halo displays may be extremely complex, especially those visible in polar regions, where air temperatures are low and the air is often full of the tiny ice crystals known as diamond dust,

which favours the formation of haloes. There are numerous different arcs and spots of light – far too many to be discussed in detail here.

The most common form of halo is the 22-degree halo, which as the name implies has a radius of 22° around the Sun. The ring is often incomplete, because of variations in the thickness of the cloud. When faint the ring appears white, but stronger haloes have a red tinge to the inner edge, sometimes grading into yellow. Very rarely, a blue-violet tint has been seen around the outer edge. The sky is darker within the halo, because light from that region is diverted away from the observer.

A second halo, with a radius of 46°, is occasionally seen. It is often incomplete, and is generally much fainter than the 22-degree halo.

Haloes may be produced by light from the Moon, but they are, of course, much fainter. Paradoxically, however, they are often easier to see than solar haloes, because of the lesser glare from the Moon. It is, however, rare for colour to be reported in lunar haloes.

Sometimes the haloes themselves are very weak and appear only fleetingly. One of the spots of light (the parhelion, discussed shortly) tends to remain visible, however, as in the photograph *(left)* of effects seen near the Moon.

Halo arcs

ID FACT FILE

Appearance:
Circles of various diameters, centred on the Sun, and arcs with both simple and complex curvature. The parhelic circle is colourless, but many arcs do show spectral colours when strong and clearly visible.

Occurrence:
Generally when a thin layer of cirrostratus covers the sky or, in polar regions, when fine ice crystals (diamond dust) are suspended in the air.

See also:
cirrostratus (p.70); circumhorizontal arc (p.139); circumzenithal arc (p.139); parhelion (p.138)

Apart from the 22-degree and 46-degree haloes just described, refraction of light through ice crystals may produce an amazing number of different haloes and arcs, so observers should be prepared for the unexpected. There are various circular haloes of differing diameters, some of which are extremely rare. Of the many other arcs, only three, the parhelic circle, the circumzenithal arc, and the circumhorizontal arc, are discussed here. Many other arcs are rare, and some vary their appearance (or existence) depending on the altitude of the Sun.

Some of the additional arcs are white, and the most common of these is the parhelic circle, which runs parallel to the horizon at the same altitude as the Sun. Under very favourable circumstances it may be followed through a full 360°. It has been known to combine with a sun pillar to produce a cross centred on the Sun.

Some of the rare halo arcs intersect the parhelic circle at various points. Generally the arcs themselves are faint, but where they cross the parhelic circle, brighter white spots may sometimes be seen. The most common are parhelia (described next), which lie near the 22° halo. These are often coloured and their tails extend along the parhelic circle. but other spots lie at 90° and 120° from the Sun, and also directly opposite it in the sky.

Parhelion

ID FACT FILE

Appearance:
Bright spots of light (sometimes with white 'tails') at the same altitude as the true Sun.

Occurrence:
Seen in any ice-crystal clouds, include ice-crystal virga.

See also:
cirrus (p.58); cirrostratus (p.70); haloes (p.132); virga (p.88)

A parhelion (more commonly known as a mock sun, false sun, or sun dog) is a bright spot of light that generally lies close to the position of the 22-degree halo, at the same altitude as the true Sun. The higher the Sun in the sky, the farther outside the 22-degree halo parhelia appear, reaching 5–6° for a solar elevation of 45°. Parhelia are not seen when the Sun's altitude is greater than 60°.

Parhelia exhibit spectral colours and are often seen in isolation (without the halo). Bright examples often have a 'tail' of white light that extends away from the Sun.

The Moon also creates similar effects. Such a mock moon is known technically as a paraselene (pl. paraselenae).

Circumzenithal and circumhorizontal arcs

ID FACT FILE

APPEARANCE:
Arcs of strong spectral colours, red closest to the Sun. The circumzenithal arc is centred on the zenith, and the circumhorizontal arc runs parallel to the horizon.

OCCURRENCE:
Seen in any ice-crystal clouds, but most often in cirrostratus and thin cirrus, with low solar elevations.

SEE ALSO:
cirrus (p.58); cirrostratus (p.70); haloes (p.132); circumhorizontal arc (p.139)

Circumzenithal and circumhorizontal arcs show some of the strongest and purest spectral colours of any optical phenomena. Only parhelia occasionally appear as strongly coloured. These arcs occur when sunlight is refracted through ice crystals, so they appear when the sky is covered by thin cirriform clouds, particularly cirrostratus and extensive cirrus.

A circumzenithal arc is a brilliant band of spectral colours, part of a circle centred on the zenith. The arc is therefore always convex towards the Sun and the colours are strongest at the nearest point, with red closest to the Sun. The arc occurs only when the Sun is fairly low in the sky (between 5 and 30° above the horizon). The radius of the circle, the strength of the colours, and the length of the arc varies with the Sun's altitude. It has a small radius, is short, and has faint colours when the Sun is low, reaches a maximum length (108°) when the Sun's altitude is 15°, and has a maximum radius, touching the 46-degree halo, for a solar elevation of 22°. The colours are generally strongest between these last two altitudes. The radius decreases and the colours in the arc tend to weaken with higher solar elevations.

Portions of an arc may often be seen in patches of ice-crystal clouds, even when other halo phenomena are absent. Circumzenithal arcs are not seen as frequently as 22-degree haloes,

but they are far more common than 46-degree haloes.

A circumhorizontal arc is a band of brilliant spectral colours, which runs parallel to the horizon below the Sun in the sky. It actually lies below the 46-degree halo, so is visible only for very high solar elevations (between approximately 58 and 80°). It cannot be seen, therefore, at high latitudes, and is invisible, except at midsummer, from middle and southern Europe and the northern USA. It may be seen farther south, including occasionally from the tropics, and from most inhabited countries in the southern hemisphere, except the southern tip of South America.

The colours are extremely striking and frequently even stronger than those seen in any circumzenithal arc. As with that arc, the length and position of the arc (and the strength

of the colours) vary with solar elevation. It touches the outside of the 46-degree halo (if the latter is visible) when the Sun's altitude is 68°. It lies farther from the 46-degree halo at greater and lesser solar elevations. Its maximum length is also approximately 108°, slightly less than one third of a complete circle around the sky.

Being produced by the same ice crystals as the circumzenithal arc, the circumhorizontal arc appears when the sky is covered in cirrostratus or moderately thin cirrus. In both arcs red is closest to the Sun. Sometimes short sections of this arc are mistaken for 'rainbows', but the latter always appear on the opposite side of the sky, and have red farthest from the Sun.

Sun pillar

ID FACT FILE

APPEARANCE:
A column of light that passes through the Sun, Moon or other source of light (such as artificial lights).

OCCURRENCE:
Seen in any ice-crystal clouds or when diamond dust is in the air.

SEE ALSO:
cirrus (p.58); cirrostratus (p.70); haloes (p.132); subsun (opposite)

A sun pillar consists of a vertical column of light passing through the Sun, the Moon or other source of light (even including artificial lights). Its colour is the same as the source, i.e., when the Sun is high in the sky it will be white, and when low, orange or red.

Sun pillars arise through reflection from flat plate or pencil-shaped ice crystals. If the whole column is visible (above and below the Sun), and it appears like an elongated figure-of-eight, then plates are responsible. If straight-sided, then pencil crystals are the cause. Light columns above artificial lights occur in cold climates when tiny crystals (diamond dust) are floating in the air.

Subsun

ID FACT FILE

APPEARANCE:
An elliptical spot of light, lying below the horizon.

OCCURRENCE:
Seen in any ice-crystal clouds.

SEE ALSO:
cirrus (p.58); cirrostratus (p.70); haloes (p.132); sun pillar (opposite)

From a high vantage point (an aircraft or high mountain), a brilliant spot of light (usually elliptical in shape) may sometimes be seen. It has the same angular position below the horizon as the Sun's elevation above it. It is actually a special form of sun pillar, caused by crystals with faces that are almost perfectly horizontal that are reflecting the light. Slight variations in the angle elongate the spot of light, but on rare occasions when the crystals are perfectly horizontal, the short column shrinks to a small spot hardly larger than the Sun.

Rainbows

ID FACT FILE

APPEARANCE:
One or more circular arcs (or portions of arcs), with radii of 46° or 51°, normally with several colours, but occasionally with only one or two colours.

OCCURRENCE:
When sunlight illuminates falling rain, on the opposite side of the sky to the Sun.

SEE ALSO:
dewbow (p.151); fogbow (p.150); sunset colours (p.162)

Most people have seen a rainbow, but few realize why it appears where and when it does, nor the range of forms that may be visible. The rainbow most commonly seen is the primary bow, and this has a radius of 46°. It is centred at the antisolar point, the point on the sky directly opposite the Sun. So the higher the Sun, the lower the top of the rainbow. This is why rainbows are not seen in the tropics during the middle of the day, but only in the morning or towards sunset. Partial bows arise when there is only a small region in which the raindrops are illuminated by the Sun.

The primary bow is created when sunlight is reflected back towards the observer (or the camera) by raindrops. This reflection occurs on the rear of the drops, which also disperse the light into its spectral colours. The order of colours is fixed: red always appears on the outside and violet on the inside of the bow.

Quite frequently, light is reflected twice within the raindrops before reaching the observer. This produces a secondary bow, with a radius of about 51° outside the primary bow. Its colours are reversed, with red on the inside and violet on the outside. Secondary bows are normally fainter than primary bows.

The area between the primary and secondary bows is noticeably darker than the rest of the sky. This dark space is known as Alexander's

dark band and arises because raindrops in that part of the sky are not returning light to the observer but instead directing it outside the field of view.

It is often possible to see additional bands of colour (known as supernumerary or interference bows) inside the primary bow. The colours are less saturated than in the primary and secondary bows and normally appear pinkish-violet or greenish in tint. They arise – like the colours seen in a patch of oil on a wet surface – through interference between rays of light that have taken slightly different paths in reaching the observer. Although normally very pale, on rare occasions the colours may be nearly as strong as those in the principal bows.

At sunrise and sunset, because short wavelengths (violet and blue) are scattered by the atmosphere, light from the Sun generally

consists of longer wavelengths (yellow, orange, and red) and just these colours appear in rainbows. Occasionally, completely red rainbows have been seen.

Very occasionally, there is a flat, reflecting surface (such as a lake) between the Sun and the observer. The reflected sunlight can produce its own bows (*p.148*). Because the light is travelling upwards from the surface, the centre of any reflected bows lies higher in the sky, but in the same general direction,

as normal bows. Generally, only a small portion of such bows is visible, often appearing as a more-or-less vertical band of colour.

Any form of spray with fairly large droplets will give rise to rainbows. They are often observed in the spray from fountains (*opposite*) and waterfalls and have precisely the same angular dimensions as those formed by raindrops. Very fine spray will produce colourless bows, like the fogbows described next.

Because rainbows are so common and have been seen by everyone, other optical phenomena are often wrongly described as 'rainbows'. They are often reported close to the Sun, for example. Although theoretically triple-

reflection rainbows can occur on the same side of the sky as the Sun, they would be extremely faint, and none has ever been reliably reported. Usually such reports are found to be misidentified parhelia (p.138) or, more rarely, to portions of a circumhorizontal arc (p.139).

The Moon also produces rainbows, although visually these are normally very faint and may appear to be colourless. Because they are caused by reflected sunlight, however, all the colours are actually present, and a time-exposure photograph could easily be taken for a daytime image.

Fogbow

ID FACT FILE

APPEARANCE:
A white arc on the opposite side of the sky to the Sun.

OCCURRENCE:
When sunlight falls on very tiny fog or cloud droplets.

SEE ALSO:
dewbow (opposite); glory (p.152); rainbow (p.144)

A white arc, known as a fogbow, with approximately the same radius (42°) as the primary rainbow, sometimes appears centred on the antisolar point. This occurs when the droplets are extremely small (as in fog, mist, or cloud), and sunlight is no longer reflected and refracted within them, but is scattered back towards the observer. Very occasionally, there may be faint blue and red tinges to the inner and outer edges, respectively.

A similar bow (a cloudbow) may be seen from an aircraft. If flying in cloud it may appear circular, but when above a cloud-bank, the bow seems to be projected onto the top of the cloud, and may be elliptical or open, with arms stretching away from the observer.

Dewbow

ID FACT FILE

Appearance:
A coloured arc apparently located on the ground on the opposite side of the sky to the Sun.

Occurrence:
When sunlight falls on droplets of dew, usually suspended from a spider's web.

See also:
fogbow (opposite); glory (p.152); rainbow (p.144)

A dewbow shows similar colours to those in a rainbow. This is most often seen in autumn, when short grass is often covered by a normally-invisible network of spiders' webs. Drops of dew hanging from the threads of the webs act just like raindrops, reflecting and dispersing the light into a spectrum.

Such a dewbow is centred on the antisolar point, as one might expect, but because the dewdrops are located on an approximately horizontal plane, a dewbow is not a circular arc, but either an ellipse, or one of the open arcs known as a parabola or hyperbola. The precise shape depends upon the elevation of the Sun and any slope of the dew-covered ground.

Glory

ID FACT FILE

APPEARANCE:
Coloured rings around the antisolar point.

OCCURRENCE:
When the observer's shadow is cast onto cloud, fog or mist.

SEE ALSO:
corona (p.128); fogbow (p.150); heiligenschein (opposite)

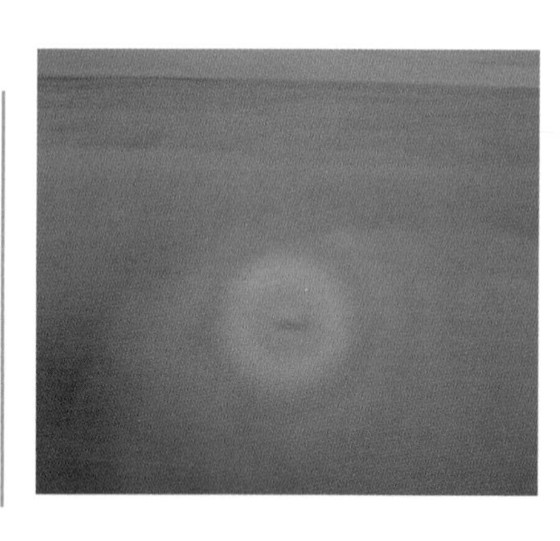

A glory is a set of coloured rings surrounding the antisolar point. The colours are similar to those in a corona, blue-violet on the inside, and red on the outer edge. Sometimes more than one set of rings is visible. Nowadays, with air travel, glories are commonly seen on lower clouds around the shadow of the aircraft in which the observer is travelling. (When the clouds lie at some distance, the shadow itself may not be apparent.) Mountain climbers may also see a glory surrounding the shadow of their head cast onto cloud or a mist- or fog-bank.

Heiligenschein

ID FACT FILE

APPEARANCE: A colourless spot of light around the shadow of the observer's head.

OCCURRENCE: When the observer's shadow falls on grass (in particular) or other vegetation.

SEE ALSO: *glory (opposite)*

A heiligenschein ('holy light') is a colourless halo that appears on dew-covered grass around the shadow of an observer's head. The dew-drops are actually held above the surface of the leaf by short hairs. Light passing through the droplet is reflected by the leaf back the way it came, so it appears bright to the observer. The mechanism is identical to that found in 'cat's eye' road markers.

A similar effect occurs on dry grass, trees, and other rough surfaces. Looking in the same direction as a ray of light, a blade of grass (for example) hides its own shadow, but looking to the side, the shadows of other blades of grass (or leaves) begin to be visible. This 'hot spot' may often be seen from an aircraft, apparently gliding over the surface of the fields and woods below.

The rising and setting sun

ID FACT FILE

APPEARANCE:
Flattening or distortions of the disk of the rising or setting Sun or Moon; distinct changes of colour.

OCCURRENCE:
Flattening of the disk is common, but major distortions occur only when there are layers of different density close to the Earth's surface.

SEE ALSO:
mirages (p.156); sky colours (p.158)

When the Sun is just above the horizon at sunrise or sunset, it is not only strongly reddened, but usually appears flattened. This is the effect of refraction. In general, the lower the layer in the atmosphere, the greater its density, and the more it refracts light. The bottom of the disk is raised by a greater amount than the top, causing it to appear flattened. When the base of the Sun's disk appears to touch the horizon, the whole disk is actually below the geometrical horizon. Refraction has 'lifted' the image into view. (The same effect naturally applies to other bodies, such as the Moon and stars.)

Very frequently, the lower atmosphere contains layers of differing densities, and these cause the outline of the Sun to appear jagged and greatly distorted. This effect is sometimes

known as 'laminated Sun'. In extreme cases, there may seem to be a reflection of part of the Sun or Moon's disk, so the body somewhat resembles a Greek letter Omega (Ω). At others, the top of the Sun appears detached, floating above the rest of the disk. Because different colours are refracted by differing amounts, occasionally the top of the disk momentarily appears in a different colour, usually green – the famous 'green flash' – and more rarely blue. Such flashes, and also a green coloration to part of the disk – the 'green segment' – are seen at both sunrise and sunset.

Mirages

Mirages occur when layers of air are significantly warmer or colder than adjacent layers. The difference in density between the layers causes light to be refracted in an abnormal manner producing inverted or displaced images of distant objects.

The most frequent type of mirage arises when the air next to the ground is much hotter than normal. The apparent pools of water on a hot road are the most common example. The 'water' is actually an image of the sky, and 'reflections in the water' are inverted images of distant objects. Such a mirage is known as an inferior mirage, because objects (e.g., the sky) appear below its true location.

When the situation is reversed, with a cold surface and a warmer higher layer of air, light from an object is refracted downwards towards the observer, and appears to come from higher than its true position. Such a superior mirage is

ID FACT FILE

APPEARANCE: Apparent inverted or reflected images of distant objects, either lower (inferior mirage) or higher (superior mirage) than their true position.

OCCURRENCE: When layers of air are significantly hotter or cooler than normal, in particular when there is a sharp contrast between adjacent layers, or between a layer and the surface.

SEE ALSO: *rising and setting sun (p.154)*

commonly seen over cold or ice-covered surfaces, such as those frequently found in polar regions.

Although mirages may be more spectacular, it is extremely common for slight distortions to occur, with distant objects appearing to be stretched or compressed, or for the horizon to appear closer or more distant than normal. Only careful observers notice these more subtle effects.

Blue sky

ID FACT FILE

APPEARANCE:
The deeper the blue of the sky, the drier and cleaner it is. Most noticeable at altitude, where the air contains little moisture and few solid particles.

OCCURRENCE:
At low altitudes, the sky is deeper in colour when the air is both dry and clean; conditions often found in the polar regions.

The blue of the sky arises because the molecules of oxygen and nitrogen in the air scatter sunlight. They strongly affect the shortest wavelengths, violet (to which humans are relatively insensitive) and blue, but have less effect on longer wavelengths (yellow, orange, and red). Blue light is scattered in all directions, which is why, to human eyes, the sky appears more-or-less uniformly blue.

Larger particles in the air, such as water molecules and fine dust, tend to scatter all the wavelengths present in sunlight, thus 'diluting' the blue that we see. This is why the sky in the mountains or seen from an aircraft appears deep blue (because the air is both dry and clean).

Aerial perspective

ID FACT FILE

APPEARANCE: Distant objects appear paler, have a blue tint, and are less distinct than nearer ones.

OCCURRENCE: Aerial perspective is always present, but it is accentuated when the air is humid, or when many solid particles are present.

SEE ALSO: *blue sky (opposite).*

At low altitudes there are often high levels of humidity and not only are there more solid particles derived from the surface (including particles of salt from the sea), but the denser air tends to keep the particles in suspension. The water molecules and the solid particles scatter white light and cause the sky to appear a pale blue.

The air at low altitudes is rarely perfectly clean and dry, so scattering causes distant objects to appear a paler bluish colour (as well as less distinct) than closer ones. There are more intervening particles to scatter the light. This effect, known as aerial perspective, is always present to some degree, and we constantly use it subconsciously to judge distances.

Haze

ID FACT FILE

APPEARANCE:
The colours of distant objects are subdued and may appear milky. Seen against the light the layer appears brown in tint.

OCCURRENCE:
Generally builds up during the day, particularly under anticyclonic conditions. Turbulent, windy conditions cause haze to disperse.

Haze consists of very small, dry particles and impedes visibility. The particles come from many different sources, both natural and man-made. They include salt crystals from the oceans, fine dust from deserts and volcanoes, carbon particles from smoke and vehicle exhausts, and innumerable pollutants. In general, these fine particles tend to scatter the shorter wavelengths, and thus appear bluish-white when seen fully illuminated by the Sun, but brownish when seen against the light. (This effect is very noticeable in the smoke from bonfires or chimneys.) The haze layer tends to be thickest at the end of the day and clearly visible at sunset.

Dust

ID FACT FILE

APPEARANCE:
Vibrant orange skies at sunrise or sunset, or distinctly blue or green disks to the Sun or Moon.

OCCURRENCE:
After major volcanic eruptions or (when there is a blue Moon or Sun) after strong winds have raised large quantities of evenly sized dust into the atmosphere.

SEE ALSO:
purple light (p.163)

Although dust is sometimes noticeable when it has been transported from a distant desert area and then deposited on the surface by rain, its presence in the upper atmosphere produces vibrant sunrises and sunsets. It is ejected to great heights by certain volcanic eruptions, together with large quantities of sulphur dioxide gas. Other particles, particularly those from forest fires, and tiny particles of soil – especially the type known as loess, found extensively in China – when lifted into the upper atmosphere, selectively absorb specific wavelengths of light and cause the disk of the Sun or Moon to appear blue or green.

Sunset & sunrise colours

ID FACT FILE

APPEARANCE:
A specific, changing, graded sequence of colours observed in the twilight arch at sunset and (in reverse order) at sunrise.

OCCURRENCE:
Best seen when the sky is completely clear.

At sunset or sunrise the sky itself goes through a sequence of colours. Above the point at which the Sun has set, the sky at the horizon is pale yellow, and the twilight arch above is bluish-white. To left and right, the horizon sky is orange, with yellow above. A little later, the twilight arch becomes salmon-pink, with yellow and orange below it. The pink arch slowly flattens and the sky above darkens from blue-grey to a purplish-blue, and finally dark blue. At the horizon, the last colours are often greenish-yellow. The sequence is reversed at sunrise and the colours are often stronger, because there is less haze.

Purple light

ID FACT FILE

APPEARANCE:
Vibrant purple skies at sunrise or sunset.

OCCURRENCE:
Days, weeks, or even months after major volcanic eruptions, depending on the time it takes for the material to spread to the observer's latitude.

SEE ALSO:
dust (p.161)

On rare occasions, when there has been a powerful volcanic eruption, volcanic dust and sulphur compounds are ejected high into the stratosphere. For some days or weeks, when conditions are correct, the twilight arch becomes a vibrant purple shade. This arises when the high-altitude particles scatter red light, which mixes with the blue always present, to give the purple hue. This colour is extremely difficult to photograph with conventional films, because of limitations on the response of most emulsions. It is readily visible to the naked eye, however, so its occurrence is immediately apparent.

Cloud colours

ID FACT FILE

APPEARANCE: Most clouds appear white unless in shadow; darker areas often indicate where clouds are evaporating.

OCCURRENCE: All clouds show changes of colour as they grow and decay; the formation of ice crystals may cause the clouds to darken slightly.

SEE ALSO: *blue sky (p.158); mamma (p.84)*

The colours of clouds during the day may vary from blindingly white to almost black. Shadowed areas normally have a bluish tint from the scattered blue light from the sky. Tiny cloud droplets scatter all wavelengths efficiently, and fully illuminated clouds appear white. Larger droplets, such as those found when clouds start to evaporate, tend to absorb light, so those areas appear darker.

Ice crystals (snowflakes) at the top of clouds give a fibrous appearance but also make the clouds seem darker. Sunlit rain falling from clouds generally appears dark, but snow and hail are much brighter.

At sunset or sunrise the clouds may change dramatically in a short space of time. Generally, at sunset, they change from grey to

yellow to red as the Sun disappears below the horizon, finally appearing dark against any higher clouds, which may remain white for a considerable time after the Sun has set. Frequently the colours are modified by scattered blue and purple light from the sky. The sequence of colours is, of course, reversed at sunrise.

The low angle of illumination by the Sun often throws cloud features into sharp relief. The most dramatic appearance is probably that shown by mamma hanging beneath the anvils of large cumulonimbus clouds, which may appear extremely menacing although, in reality, they are not signs of violent weather.

Crepuscular rays

ID FACT FILE

APPEARANCE: Shafts of light or shadows cast by clouds, hills or mountains onto the atmosphere itself.

OCCURRENCE: When the atmosphere is sufficiently hazy or humid for the rays to be seen, or when there are gaps in a layer cloud.

SEE ALSO: *mountain shadow (p.169); shadow of the Earth (p.171); stratocumulus (p.40)*

Occasionally, rays of light and intervening shadows may be seen radiating from the Sun. Frequently seen when the Sun is low at sunrise or sunset, they are known as crepuscular rays. Similar effects may, however, be seen at other times of the day, so three, slightly different types, are recognized:

- shadows cast by clouds, hills or mountains when the Sun is low (or even below the horizon) and the intervening bright rays
- shadows and bright rays produced around the edges of clouds
- shafts of light penetrating through gaps in the cloud cover

The first type, seen at twilight – which was the source of the name 'crepuscular' – are sometimes extremely long. They have been known to appear even though the clouds

causing them were below the observer's horizon. Similarly, they occasionally extend right across the sky and seem to converge at the antisolar point. Such converging rays are sometimes called 'anticrepuscular rays'.

The second form is quite common, and occasionally the edge of a cloud will seem to be outlined in shadow. This type is particularly strong when the air is hazy or very humid. When the atmosphere is completely clear and free from any particles, all the different types of crepuscular rays are less prominent.

The third type frequently occurs with broken cloud, such as stratocumulus. The shafts of light are sometimes described as 'the Sun drawing water' although, of course, no such thing is happening. An old nautical term is

'Apollo's backstays', and they are also sometimes called 'Jacob's ladder'.

Quite frequently, the darker, shadowed areas between these crepuscular rays are mistaken for shafts of rain, especially when the clouds and gaps are at a considerable distance from the observer. Identification of the cloud type – which is often stratocumulus – will soon correct this impression. Stratocumulus, in particular, never gives rise to any significant rain.

Mountain shadow

ID FACT FILE

APPEARANCE: A dark cone stretching away from the observer.

OCCURRENCE: When a mountain top is illuminated by the rising or setting Sun.

SEE ALSO: *alpine glow (p.170); crepuscular rays (p.166); shadow of the Earth (p.171)*

To an observer standing on a high hill or mountain at sunrise or sunset, the shadow of the peak appears as a dark cone stretching away into the distance. The sides of the shadow are actually parallel to one another, but appear to converge through perspective.

The clearest mountain shadows occur when the shadow falls on a layer of lower cloud. If the shadow (like crepuscular rays) is actually cast on the atmosphere, which happens to be hazy, the outline is slightly less distinct.

Alpine glow

ID FACT FILE

APPEARANCE:
A sequence of colours from yellow to purple seen on mountain-tops at sunset and (in reverse order) at sunrise.

OCCURRENCE:
Only visible when there are clear skies to both eastern and western horizons.

SEE ALSO:
shadow of the Earth (opposite)

At sunrise or sunset, the tops of mountains (or high cumulonimbus clouds) are sometimes illuminated by a striking sequence of colours, known as the alpine glow. With clear skies at sunset, the light on the peaks may become, in turn, yellow, orange, red, and finally purple. The last colour actually results from the mixing of red light from the Sun with weak scattered blue light from the sky. The effect is accentuated if the peaks are covered in snow and ice. The sequence is, of course, reversed at sunrise.

Shadow of the earth

ID FACT FILE

APPEARANCE:
A blue-grey segment, often with a reddish upper edge, that rises in the east at sunset, or falls in the west at sunrise.

OCCURRENCE:
When the sky is clear both to the east and west at sunrise or sunset.

SEE ALSO:
alpine glow (opposite); mountain shadow (p.169)

If the air is clear as the Sun sinks in the west at sunset, the shadow of the Earth itself, cast onto the atmosphere, may be seen rising in the east. It has a steely-grey colour and usually has a reddish upper border. As the shadow rises, it darkens and finally merges with the dark sky overhead. The effect is most pronounced when seen in clear mountain air, but is often observable (but usually overlooked) from the lowlands. A similar, reversed sequence naturally occurs at sunrise.

Rain

When air reaches its dewpoint, cloud droplets form on innumerable tiny condensation nuclei, which are present throughout the troposphere. Cloud droplets are so small (0.001–0.05 mm in diameter) that they easily remain suspended in the atmosphere and growth by condensation or collision is very slow. Raindrops are much larger (about 0.5–2.5 mm in diameter), and there are two processes by which they form, giving rise to what are known colloquially to meteorologists as 'warm rain' and 'cold rain'.

Several processes combine to produce warm rain, but primarily the largest drops fall faster than smaller ones, overtake and collide with them, gradually growing larger (and falling faster). This basic process occurs most easily in very deep cumulus congestus clouds, as found in the tropics at all times of the year and in summer in temperate latitudes. Shallow cumulus clouds, or stratiform ones such as stratus and stratocumulus do not produce large raindrops or great quantities of rain, but St and Sc may often give rise to fine drizzle. It has been calculated that it takes between 20 and 60 minutes for an average-sized raindrop to form, so short-lived clouds are unlikely to produce any rain.

Cold rain is created by a completely different process which involves the glaciation (freezing) of cloud droplets and therefore occurs most commonly in cumulonimbus clouds. Even here, however, it is not sufficient for the temperature to drop below 0°C. Just as condensation requires condensation nuclei, freezing requires similar nuclei. If no suitable nuclei are present, cloud droplets can continue to exist as liquid water as low as -40°C (when they are said to be supercooled), before they freeze spontaneously. Supercooled droplets are found in many clouds, and also in certain fogs, and they freeze instantaneously as soon as they come in contact with a suitable nucleus or a solid surface (such as an aircraft).

As might be expected, the size of ice crystals that form depends on the exact temperature. At low temperatures (about -30°C) many tiny crystals are created, but at higher temperatures (about -10°C), a relatively small number of large crystals are formed. When the crystals fall into warmer air at lower altitudes they naturally melt. Tiny droplets often evaporate before reaching the ground, but larger droplets survive to fall as rain. The largest raindrops are about 2.5 mm in diameter. Drops that grow larger than this size tend to split into two or more smaller droplets.

In summer, even the tops of high cumulus congestus clouds may not reach levels at which glaciation takes place. In winter, by contrast, quite shallow cumuliform clouds may become glaciated, and produce a shower of rain. This is why the transition from cumulus congestus to cumulonimbus calvus is a crucial stage in the development of showers, and one that observers need to be able to recognize.

Rain and drizzle

By convention, rainfall is divided into rain and drizzle:

- rain: drops with diameters greater than 0.5 mm (up to a maximum of about 2.5 mm)
- drizzle: droplets with diameters below 0.5 mm

These two types of precipitation are produced by different clouds:

- rain: cumulonimbus, cumulus congestus, altocumulus, altocumulus floccus, altostratus, nimbostratus, thick stratocumulus (Sc str op)
- drizzle: stratus, thick stratocumulus (Sc str op)

When stratus and stratocumulus appear to bring rain, it is usually because they have been seeded by ice crystals falling from a higher cloud layer (such as altostratus).

Rain from cumulus congestus and cumulonimbus occurs in the form of showers. Generally, the rain from cumulus and many small cumulonimbus clouds although often heavy, is restricted in area (above) as well as in duration. Only rarely will the swathe of countryside affected by the rain be more than a few kilometres in width. With larger cumulonimbus clusters, there may, of course, be several cells producing rainfall, so a larger area is affected. Even so, some areas may escape rain altogether.

This is in complete contrast to the rain at frontal systems, where the main rain-bearing clouds are altostratus,

nimbostratus (above) and, to a lesser extent, thick stratocumulus. Here the area affected may amount to thousands of square kilometres. In such cases the heavy rain tends to be arranged in bands roughly parallel to the surface fronts, with areas of lesser, or no, rainfall between the bands. Cumuliform clouds are sometimes embedded in the normally stratiform clouds, and may then produce bursts of intense rain or hail.

Occassionally rain or drizzle falls onto the ground or other surfaces that are at extremely low temperatures, well below freezing. The droplets freeze instantly, covering the surface with tiny hemispherical beads of ice. Known as freezing rain or freezing drizzle, this occurs only when precipitation is light. Heavier rain would spread out to produce a layer of liquid water on any surface rather than individual droplets. When such a layer freezes it is known as glaze.

Snow

If ice crystals survive to reach the ground without melting, the result is a fall of snow. Individual crystals form with the beautiful, complex hexagonal shapes that are commonly regarded as 'snowflakes'. When the temperature in the snow-bearing clouds is just below 0°C, many of the crystals have a thin surface film of water. This freezes when crystals come into contact, locking them together. It is these aggregates of many individual crystals, rather than individual ice crystals that form the snowflakes that float down to the ground. At very low temperatures, such aggregates do not form, and the tiny crystals fall as powder snow, much to the delight of skiers. Such dry snow is also favoured by transport authorities, because it may be cleared by the use of snow blowers, whereas the wet snow created close to freezing point melts with the slightest pressure, but refreezes as soon as the pressure is released. Wet snow readily becomes compacted into a sheet of ice on roads and railway lines, creating hazardous conditions and disrupting traffic, as well as causing severe problems at airports.

Blowing snow may pose particular hazards. With high winds, blizzard conditions may not only reduce visibility, but also create major drifts that block roads or railways. In areas where snowfall is frequent, snow fences are often constructed some distance away from vulnerable stretches of road or railway. The obstruction to the wind creates eddies on its leeward side, where the snow accumulates as drifts well away from the road or railway track.

Freshly fallen snow contains a lot of air, but after a while the snow becomes compacted and begins to change its nature. Larger crystals grow at the expense of smaller ones, and this may create a weaker layer below the surface. Such conditions are likely to give rise to avalanches, where the uppermost layers of snow become detached and slide over the weaker layer beneath. This is often triggered by vibrations (either natural or created by human activity) but may also occur when a thaw sets in or there is heavy rainfall.

Snow that has lain for more than a year is known as firn, as are the granular layers beneath the surface of a snow-pack. If the snow persists for several years it eventually becomes compacted into glacier ice.

Hail

A hailstone consists of alternating layers of clear and opaque ice. These layers are laid down in regions of a cloud that are at different temperatures. The clear layers are deposited in warm regions where liquid droplets spread over the surface of the hailstone before freezing. The opaque layers are laid down where the cloud droplets are supercooled, freezing instantly on contact with the growing hailstone, trapping air between them. The conditions for this process to occur are found in deep, vigorous cumulonimbus clouds with strong updraughts that are able to support ice pellets and lift them through the different layers. Generally such updraughts are tilted, so that when the hailstones are carried aloft and are tossed out of the rising air, they fall back down towards the surface, but encounter the updraught again at a lower level and are carried upwards again. This may happen several

times, adding multiple layers of ice to the hailstone, until it becomes too heavy and falls out of the cloud.

The largest hailstones are produced in supercell storms, which have particularly strong updraughts. Some may reach the size of grapefruit and weigh 1 kg. Even larger hailstone aggregates are known where several individual hailstones have become frozen together, reaching as much as 4 kg. Such stones may be devastatingly destructive, but even the smaller hail produced by multicell storms or individual cumulonimbus clouds may cause considerable damage. Hailshafts usually appear distinctly white in comparison with greyish falling rain. The photo *(opposite)* shows that both hailshafts and rain are falling at the same time. (The presence of rain is confirmed by the rainbow.)

Dew

Dew is deposited when, normally under clear skies, objects rapidly lose their warmth after sunset, cooling below the dewpoint. Generally the water vapour comes from the ground, rather than the air, because the soil remains slightly warmer than the blades of grass, leaves and other objects. If the temperature continues to fall, spiders' webs and other objects may become covered in frozen dew.

Dewdrops are small – usually less than 1 mm in diameter. When they occur on grass (in particular) they may produce the optical effect known as the heiligenschein (p.153) when illuminated by sunlight, and occasionally a dewbow (p.151). The large drops sometimes seen on the very tips of blades of grass or leaves are not dew. Plants transport large quantities of moisture from the soil through their leaves, where it normally evaporates into the air. When the humidity is high, it cannot evaporate, but instead oozes from the tips of the leaves to form what are called guttation drops.

Hoar frost

Hoar frost is very common. It consists of water vapour that is deposited directly onto objects as soft crystals of ice, rather than as liquid water that then freezes. Generally crystals appear first along the edges of objects and at the tips of leaves, where temperatures are lowest.

When dewdrops first occur on objects, and the temperature continues to fall, the droplets may not freeze immediately the temperature reaches 0°C. They become supercooled (p.172). Once a few ice crystals do occur, these grow rapidly at the expense of neighbouring supercooled droplets. This is the origin of the beautiful patterns often seen on window-panes, where close examination will often show that there is a narrow gap between the ice crystals and the still liquid supercooled droplets.

Rime

A coating of rime on exposed objects is often taken for hoar frost, but it is produced by a completely different process. It is deposited from supercooled fog. As the fog drifts across the ground, the droplets freeze almost instantaneously when they touch solid objects. A hard coating of rough ice builds up on windward surfaces and the 'feathers' point into the wind. Even in lowland areas large deposits may occur if the fog persists, but in mountainous regions the masses of ice may be as much as several metres across.

Although rime's 'feathers' normally point into the wind, under very still conditions, long, needle-like crystals may grow on the very edges of leaves and other objects. These are much longer than the crystals normally found with ordinary hoar frost.

Glaze

Glaze is a layer of more-or-less clear ice that coats objects on the ground. It forms when supercooled raindrops or drizzle come into contact with cold objects whose temperatures are below freezing. The droplets do not freeze instantly, but have time to spread out into a thin layer before they are converted to ice.

The 'black ice' often mentioned in weather forecasts is more correctly known as glaze and it does, of course, pose a particular hazard, because it is clear and not readily visible. Large accumulations of glaze may occur when rain from an approaching warm front falls into a layer of frigid air at the surface. Such 'ice storms', like the one that hit Canada and parts of the United States in 1999, may be extremely destructive, when the glaze snaps trees, utility poles and electricity pylons.

Mist and fog

Although technically they are not precipitation, mist and fog may be conveniently dealt with here. Both consist of fine water droplets suspended in the air; mist having a visibility greater than 1 km, and fog a visibility less than 1 km. Both occur when air is cooled to its dewpoint as happens in clouds, and both may be regarded as cloud at ground level.

There are two common forms of mist or fog:

- radiation fog: produced when the ground cools rapidly (usually at night), chilling the air above it
- advection fog: produced when a wind carries humid air over a cold surface, which cools the air, forming fog, which then spreads to adjacent areas

Radiation fog is likely to form when the sky is clear (allowing heat to radiate away to space), the air is humid, and there is

sufficient time for the temperature to fall to the dewpoint. Such conditions are most frequent in autumn and early winter, especially after a fall of light rain. In addition, the wind should be light (no greater than about 7 kph), otherwise turbulence mixes the cold air near the ground with warmer air from a higher layer, and prevents the temperature from falling very low. Such conditions are often found in valleys, which is why valley fog is so common.

Advection fog commonly forms over the sea or large lakes when warm, humid air flows above cold water. The resulting fog is then carried inland by the wind, although if this is too strong, it tends to produce low cloud, rather than fog.

A rare form of fog is known as steam fog. It occurs when very cold air flows above warm water. Water vapour evaporates from the sea, producing wisps that look like steam. It is particularly common in polar regions, where it is known as arctic sea smoke.

Winds

Winds are always described by the direction from which they are blowing. Thus the dominant westerlies blow from west to east, and the trade winds (for example) are north-easterlies in the northern hemisphere, and southeasterlies in the southern. Two terms are used to describe a change in wind direction, and these are the same in both hemispheres:

- veer: wind direction changes clockwise, e.g. south to southwest
- back: wind direction changes anticlockwise, e.g. southwest to south

There are three important factors that determine the strength and direction of the wind. These are:

- the difference in pressure between two points (known as the pressure gradient), usually created by differences in temperature
- the rotation of the Earth
- the degree of friction created by the surface (p.191)

Although the wind is the result of the combined action of these factors, they may be considered separately. The first two are extremely significant in the global circulation, which is the source of weather systems.

The global circulation

The global circulation is driven by the imbalance between heating in the tropics and cooling at the poles. Because of the Earth's rotation, the circulation does not form one large cell in each hemisphere. Instead, three smaller cells exist. Warm air that rises in the tropics descends in the subtropical anticyclones (high-pressure areas), centred approximately at 30° N and S. From here, air at the surface flows both back towards the tropics – creating the trade winds – and also towards the poles. Because of the Coriolis effect, the winds do not flow directly north and south, but curve to the right

The Coriolis effect

Because of the rotation of the Earth, air does not flow directly from high pressure to low. A parcel of air at the equator, for example, is carried eastwards very rapidly by the Earth's rotation. If it moves north (say) its eastward motion is greater than the underlying surface, so the parcel of air curves towards the right. A similar situation occurs for polar air moving towards the equator. In the southern hemisphere the curvature is towards the left. So winds veer (turn clockwise) in the northern hemisphere, and back (turn anticlockwise) in the southern.

This Coriolis effect is responsible for the global wind patterns. Air flowing from the mid-latitude high-pressure zones towards the equator creates the trade winds, northeasterly in the northern hemisphere, and southeasterly in the southern. Air flowing towards the poles from the same high-pressure zones creates the dominant westerlies over temperate latitudes. Finally, air flowing from the poles towards the middle latitudes becomes the polar easterlies.

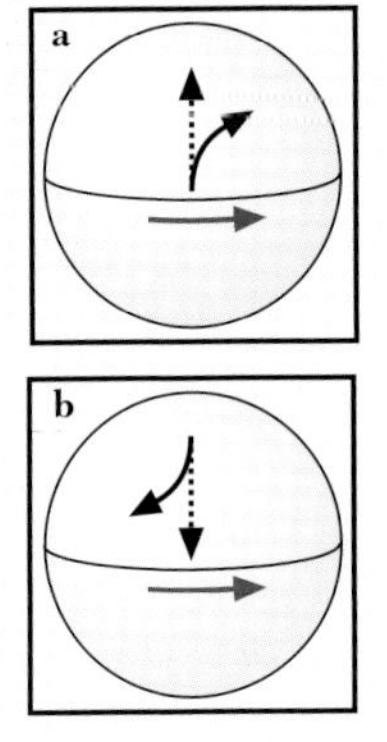

in the northern hemisphere and to the left in the southern. Belts of mainly westerly winds lie on the poleward side of the subtropical anticyclones, approximately between latitudes 40 and 70° N and S, although (particularly in the northern hemisphere) they vary greatly in strength and position. These westerlies are all-important for the weather in temperate regions. The prevailing winds in these zones are from the west or southwest in the northern hemisphere, and west or northwest in the southern hemisphere. These steer weather systems eastwards around the globe. Although winds encircle depressions and anticyclones, the systems

themselves (and particularly depressions) move from west to east, following the general flow.

Cold, dense air flows from the poles towards the equator, giving predominantly easterly winds at the surface. An extremely important atmospheric boundary occurs where the polar easterlies meet the warmer westerlies that encircle the globe at middle latitudes. A sharp temperature contrast exists across this boundary, known as the Polar Front. (Fronts are described in more detail shortly.) The two polar fronts (one in each hemisphere) exhibit a series of irregular and unstable waves around the globe (usually four or five), where warm moist air extends towards the poles, and colder air towards the equator. Depressions are created at the polar fronts and, by mixing polar and tropical air, transfer the warmth of the tropics towards the poles.

Semi-permanent features

In the tropics, the hot, moist air creates active cumulus congestus and cumulonimbus clouds, which produce abundant rain. Heavy thunderstorms are a daily occurrence in many parts of the tropics. Satellite images generally show a line of major convective clouds where the northeast and southeast trade winds converge along a boundary known as the Intertropical Convergence Zone (ITCZ).

As air descends in the subtropical anticyclones it is compressed and warmed, producing clear skies and intense solar radiation. In the northern hemisphere, for example, the Sahara and Arabian deserts, and the southwest of the United States are often completely cloud-free.

At temperate latitudes, cloud cover varies greatly, because the weather is dominated by a succession of depressions, interrupted by extensions of the subtropical anticyclones or other anticyclones with relatively clear skies.

The cold polar easterlies spread down to about latitudes 60° N and S, where warming from the oceans produces semi-permanent low-pressure areas, the most important being the winter Icelandic and Aleutian Lows and the Siberian High in the northern hemisphere. The last of these decays in summer to become the Asian Low, and the alternation between these two states drives the Asian monsoons.

Jet streams

As mentioned earlier, the tropopause, the upper boundary of the troposphere, is highest over the tropics and lowest over the poles. There are generally distinct breaks in its level above the subtropical anticyclones and also near the polar fronts. At these breaks, the great temperature contrasts generate the subtropical and polar-front jet streams. The sub-tropical jet, marked by well-developed cirrus clouds, is seen here crossing the Nile Valley and the Red Sea. These fast-flowing ribbons of air – which may reach 400 kph or more – snake around the Earth, like the polar fronts, but are very variable in strength and may split or disappear over part of their course. They (and the polar-front jet in particular) have a great effect on the growth, movement and decay of weather systems below them.

Upper and lower winds

One result of the Coriolis force is that winds do not, as one might expect, follow the pressure gradient, flowing from high pressure to low. Instead – away from the surface – they follow what is known as Buys-Ballot Law:

- facing downwind, low pressure is to the left in the northern hemisphere (to the right in the southern).

This law applies to freely flowing air at altitudes above about 500–600 m, (known technically as the geostrophic wind). This wind, therefore, flows around high- and low-pressure areas. With some exceptions for local winds, which we discuss later, the lowest clouds will follow this geostrophic wind.

The wind at the surface, however, will be different in both strength and direction, because of friction. The lowest layer of air is slowed, and this reduces the Coriolis force. As a result, the wind no longer flows round the high- and low-pressure areas. Instead, it flows out from a high-pressure region, and in towards the centre of a low. Relative to the geostrophic wind, the surface wind is backed in the northern hemisphere, veered in the southern.

The amount of change in the wind direction depends upon the nature of the surface. Over the sea, the change is about 10–20°, whereas over the land it may amount to 30–50°. So the Buys-Ballot Law must be modified for the surface wind. Instead of the low-pressure centre being directly to the left (in the northern hemisphere) it is between 10 and 50 degrees farther forward. This is an important point to remember, because otherwise it is easy to be deceived by the wind direction. When directly facing the surface wind and more-or-less at sea level, for example, clouds immediately in front of you will not move overhead. On land, it is a shower, 30-odd degrees to the right of the surface wind, that is likely to drench you with rain.

Naturally, the situation changes if you are up in the mountains. On the windward side of a high mountain, the wind is likely to be similar to the geostrophic wind, but on the leeward side of a rugged mountain range, the wind is likely to be closer to the wind observed in the plains. (This ignores funnelling and eddy effects, of course.)

Mountains may also completely block the airflow. In the illustration (*below*) a northwesterly wind has been deflected to flow eastwards along the northern edge of the Pyrenees. Cloud has been held back to the northern side of the range.

Wind speeds

Meteorologists measure wind speeds on land at a standard height of 10 m, but may use other heights at sea (on moored buoys, for example). Speeds are normally quoted in metres per second, or knots, the latter being standard in aviation. For general use (and in this book), speeds are often quoted in kilometres per hour, which are easier to relate to everyday experience.

Weather forecasts commonly describe winds in terms of the Beaufort Scale. Originally defined in terms of the sails that could be carried by naval vessels, it has been redefined in general terms for modern usage, and also extended to provide a scale for use on land.

Beaufort Scale for use at sea

Force	Description	Sea state	knots	kph
0	Calm	Like a mirror	below 1	below 2
1	Light air	Ripples; no foam	1–3	2–6
2	Light breeze	Small wavelets with smooth crests	4–6	7–11
3	Gentle breeze	Large wavelets; some crests break; a few white horses	7–10	12–19
4	Moderate breeze	Small waves; frequent white horses	11–16	20–30
5	Fresh breeze	Moderate, fairly long waves; many white horses; some spray	17–21	31–39
6	Strong breeze	Some large waves; extensive white foaming crests; some spray	22–27	40–50
7	Near gale	Sea heaping up; streaks of foam blowing in the wind	28–33	51–61
8	Gale	Fairly long & high waves; crests breaking into spindrift; foam in long prominent streaks	34–40	62–74
9	Strong gale	High waves; dense foam in wind; wave-crests topple and roll over; spray interferes with visibility	41–47	75–87
10	Storm	Very high waves with overhanging crests; dense blowing foam, sea appears white; heavy tumbling sea; poor visibility	48–55	88–102
11	Violent storm	Exceptionally high waves may hide small ships; sea covered in long, white patches of foam; waves blown into froth; visibility severely affected	56–63	103–117
12	Hurricane	Air filled with foam and spray; extremely bad visibility	≥ 64	≥ 118

Beaufort Scale for use on land

Force	Description	Events on land	knots	kph
0	Calm	Smoke rises vertically	below 1	below 2
1	Light air	Direction of wind shown by smoke, but not by wind-vane	1–3	2–6
2	Light breeze	Wind felt on face; leaves rustle; wind-vane turns to wind	4–6	7–11
3	Gentle breeze	Leaves and small twigs in motion; wind extends small flags	7–10	12–19
4	Moderate breeze	Wind raises dust and loose paper; small branches move	11–16	20–30
5	Fresh breeze	Small leafy trees start to sway; wavelets with crests on inland waters	17–21	31–39
6	Strong breeze	Large branches in motion; whistling in telephone wires; difficult to use umbrellas	22–27	40–50
7	Near gale	Whole trees in motion; difficult to walk against wind	28–33	51–61
8	Gale	Twigs break from trees; difficult to walk	34–40	62–74
9	Strong gale	Slight structural damage to buildings; chimney pots, tiles, and aerials removed	41–47	75–87
10	Storm	Trees uprooted; considerable damage to buildings	48–55	88–102
11	Violent storm	Widespread damage to all types of building	56–63	103–117
12	Hurricane	Widespread destruction; only specially constructed buildings survive	≥ 64	≥ 118

Local winds

Apart from the winds that arise from the general distribution of high- and low-pressure areas around the globe, there are other winds that have more localized sources. The most important of these are:

- sea and land breezes
- valley and mountain winds
- lake breezes
- föhn winds
- fall winds

The strength and direction of the wind may also be affected by hills, valleys, cliffs, mountains and other features, through funnelling effects, the formation of eddies, and by presenting a physical barrier that the wind must either flow round or over. These are not discussed in detail here.

Sea breezes

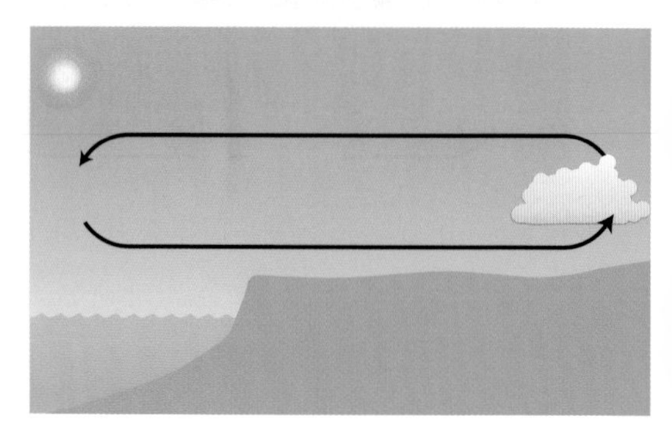

During daylight, the land heats rapidly in sunlight (especially in summer), while the sea remains cold. Air rising from the warm land draws cooler air in from the sea, creating a sea breeze. This usually begins around local noon, and is strongest during the afternoon, before dying away at sunset. A shallow circulation arises, with a return flow at low altitude. In spring and early summer, the sea breeze often brings a sea mist in from the coast. There is usually a distinct sea-breeze front, where there is a distinct change in temperature. The front penetrates farther and farther inland as the day progresses, and is often marked by a line of convective cloud (*opposite*), which is sometimes active enough to give rise to showers, especially if there is a range of hills inland to provide additional uplift.

Sea breezes occur on all scales, from very localized ones to others that sweep in from coasts hundreds of kilometres long. They tend to blow at right-angles to the coast. If the land is in the form of a peninsula, sea breezes often sweep inland from both coasts until their fronts combine to produce a line of convective cloud along the centre of the peninsula.

Land breezes

Land breezes are the night-time counterpart of sea breezes. They arise because the land (particularly on clear nights) cools much more rapidly than the sea. At about midnight a breeze begins to blow from the land out to sea, and this generally continues until dawn. As with sea breezes, there is a land-breeze front, often marked by a line of cloud, which may be particularly distinct on early-morning satellite images.

Valley winds

Valley winds arise when hill- or mountain-sides are heated by the Sun. The warm air rises, and a wind starts to blow either along its length towards its head, or up one of its sides towards the ridge. If the air is humid and lifted above its condensation level, cloud forms at the head of the valley or along the ridge. A similar process sometimes occurs when the valley has been filled with shallow overnight fog. The warmth of sunlight lifts the fog, which usually breaks up into patches of low stratus. These then drift up the valley and the mountain-sides, sometimes producing a line of stratus along one slope which slowly dissipates.

The exact nature of a valley wind depends upon the valley's orientation, but the flow of air usually starts shortly after sunrise, and may continue all day, reaching a maximum of 15–20 kph, and dying away at sunset. In valleys that are shaded from sunlight, there may be a slight drift of air during the day, but generally this will be masked by any overall pressure-gradient wind.

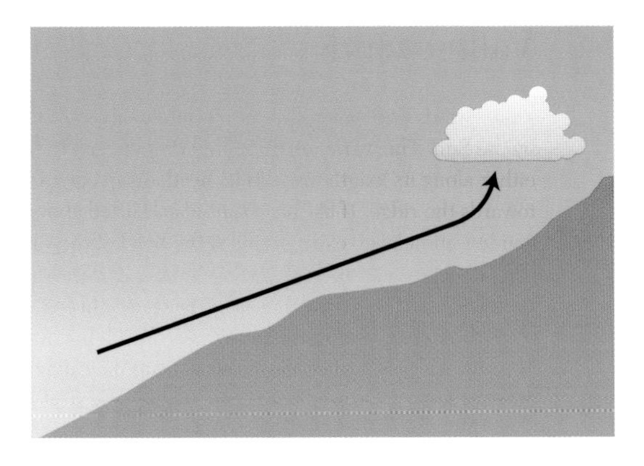

Mountain winds

Mountain winds are the night-time counterpart of valley winds. At night, the upper slopes of mountains and valleys cool more quickly than the more sheltered valleys by radiating heat away to space. The air in contact with them is cooled, becomes denser, and slides downhill. Such winds begin shortly after sunset, and continue until shortly after dawn. If several valleys drain into one larger one, there may be a considerable flow of air down towards the plains, but wind-speeds tend to be less than the corresponding valley winds – up to about 12 kph – although the wind may be greater if it is forced into a narrow canyon or gorge.

These mountain winds, which are created by the night-time cooling of mountain and hill slopes, should not be confused with fall winds (p.201) that arise over snow- or ice-covered high ground, and which are frequently far more violent. Valley and mountain winds, like sea and land breezes, are part of a simple, and relatively gentle, day/night cycle.

Lake breezes

Large areas of deep water, such as the Great Lakes of North America, may produce their own systems of winds, similar to sea and land breezes, and caused in exactly the same way. Shallow lakes warm rapidly during the day, and cool rapidly at night so there is less of a temperature contrast, and any wind may be weak and likely to die away. Small lakes do create lake breezes, however, if their effects are amplified by surrounding mountains, which, when heated by the Sun, produce their own breezes – equivalent to valley winds – during the day.

Föhn winds

When air rises over mountains, it often produces precipitation on the windward side, but when it starts to descend to leeward, it has lost much of its moisture. Just as air cools when it rises, it warms when it descends, but dry air warms faster than saturated air. When air is drawn across a mountain range, the air to leeward may be much warmer and drier than at the same level on the windward side. Such winds are known as föhn winds, from the term used in the Alps. The sudden arrival of a föhn may cause a dramatic rise in temperature, which may be enough to create a sudden thaw and rapid melting of any snow cover. The North-American Chinook is a föhn wind that descends from the Rockies.

Persistent föhn winds may create a considerable fire hazard in areas with wooden buildings, because timber tends to become dessicated and therefore burns readily. They are also well-known for their effects upon humans, who tend to become short-tempered and aggressive, as well as suffering from aches and pains, caused by the extremely low humidity and increased temperature.

Fall winds

Fall winds arise when cold air that has accumulated over high land (such as a plateau) cascades downhill, sometimes assisted by an overall pressure gradient. Despite warming during its descent, such winds usually remain cold. They may be extremely violent, especially those around Antarctica, where Commonwealth Bay, 'The Home of the Blizzard', has recorded speeds of 320 kph. The mistral that blows down the Rhône valley and over the Golfe du Lion is a well-known fall wind.

Named winds

There are many names for specific local winds, which may be of a particular type or occur at certain times of the year. A few well-known ones are given here.

Wind	Location	Type
Bora	Adriatic	Fall wind
Chinook	N. America	Föhn wind
Etesian	Aegean	Seasonal northerly
Gregale	W. Mediterranean	Northeasterly
Harmattan	NW. Africa	Dry easterly or north easterly
Levanter	E. Spanish Coast	Humid easterly
Meltemi	Aegean	Seasonal northerly
Mistral	Rhône valley	Fall wind
Santa Ana	S. California	Föhn wind
Sirocco	Mediterranean	Warm southerly
zonda	Argentina	Westerly föhn

Air masses

Air masses are an extremely important factor in determining the weather. An air mass is a large body of air, normally covering several thousand square kilometres, that acquires specific characteristics (temperature and humidity) because it is stationary over a particular part of the globe for a long time. When it moves from its source region it initially retains its characteristics, but these may become modified with time, depending on the surface over which it moves. In general terms, temperature is determined by where the air mass originated, and humidity by whether it began, or has passed over, a continental or oceanic surface. The principal sources are the polar regions, the sub-tropical highs, and the equatorial zone. In the northern hemisphere, important sources are the Arctic, the Siberian (winter) High, the Azores (Bermuda) High over the Atlantic, and the Pacific High. In the southern hemisphere, source regions are the dominant Antarctic High, the South Atlantic and South Pacific Highs, and the winter South Australian High.

Four major classes are recognized:

- A Arctic and Antarctic (the latter sometimes designated AA)
- P Polar
- T Tropical
- E Equatorial

Any air mass that originates or travels over the sea tends to become humid, and is defined as being 'maritime' (m), whereas one that originates in the centre of a continent and has not travelled far over water remains 'continental' (c).

The major types are:

- mA maritime Arctic extremely cold and humid
- mP maritime Polar cold and humid
- cP continental Polar cold and dry
- mT maritime Tropical warm and humid
- cT continental Tropical hot and dry
- mE Equatorial hot and humid

Arctic, and particularly Antarctic, air is always exceptionally cold and dry. (The humidity of any extremely cold air is generally very low, regardless of whether it forms over land or ice-covered seas).

The cloud cover and weather are strongly determined by the nature of the air mass and the surface below it. Warm, humid maritime tropical air that moves towards the pole generally moves over colder waters. The lower levels are cooled, producing stable conditions, with sea fog, and low stratiform cloud (stratus or stratocumulus). Cold maritime polar air (or the even colder Arctic air) that moves towards the equator across warmer seas is warmed at its base, and becomes very unstable, giving rise to numerous cumulonimbus clouds and showers. Stable air often occurs in the warm sector of depressions and instability behind cold fronts.

Fronts

The boundary between two air masses with different characteristics is known as a front. Because there is always a difference in temperature and humidity between the two air masses, the densities also differ. One air mass (the colder and denser) tends to undercut the other (warmer and lighter), lifting it away from the surface. Occasionally a front will remain in a quasi-stable stationary state, but any small disturbance destabilizes the situation, causing both air masses to move.

Such a situation is common at the Polar Front in both hemispheres, which is where depressions (low-pressure areas) form. Initially, the Front may be quasi-stationary, with the cold polar easterlies on one side, and warm tropical westerlies on the other. This state does not persist for long, and a wave begins to develop. On the eastern side, the cooler air begins to retreat slowly back towards the pole, and the warm tropical air advances and starts to slide up over the lower layer. On the western side of the wave, the cold polar air starts to advance towards the equator, undercutting the warm air as it does so. The satellite image shows a series of waves on the Polar Front, just to the west of the British Isles.

The warm front (*left*), where the warm air is advancing, has a fairly shallow slope of about 1:100 to 1:150. The wedge of warm tropical air slides up over the cooler polar air, which is slowly forced to retreat. The cold front (*right*), where the cold air is undercutting the warm air, tends to be steeper

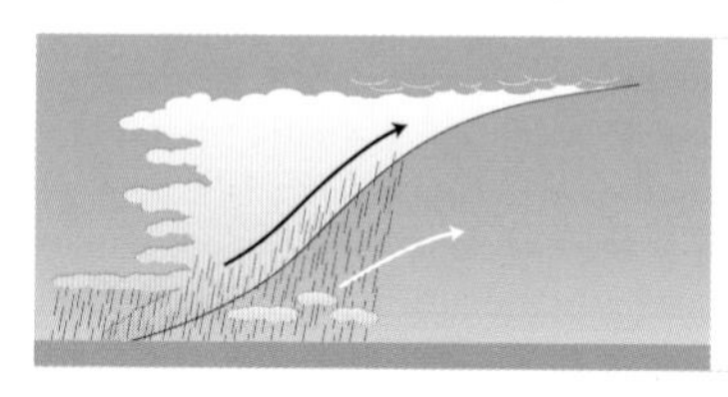

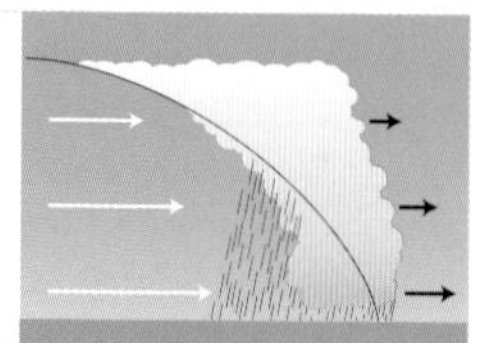

(about 1:50 to 1:75) and advances more rapidly across the surface. A third type of front (an occluded front) will be described shortly. All three types are shown by specific symbols on weather charts.

Fronts may, of course, arise away from the main Polar Front. Cold fronts, in particular, may advance across country as a line squall, with violent convective activity in the form of thunderstorms (often accompanied by heavy hail).

Pressure

Pressure is measured by barometers of various types, and in most countries (the main exception being the United States) pressure values are given in millibars (mb), the average sea-level pressure being 1013 mb. Meteorologists have switched to using a unit known as the hectopascal (hPa), which is more rigorously defined scientifically. However, one hectopascal is precisely equal to one millibar, which is the unit best understood by the general public.

Mapping the weather

A vital guide to current and future weather is a pressure chart – known techincally as an isobaric chart. Points with equal barometric pressures are joined by lines known as isobars. Although meteorologists use such charts plotted for various heights in the atmosphere, the type most commonly encountered, and which is found in television broadcasts and newspapers, shows the pressure at sea level. Observations of pressure from reporting stations are adjusted to compensate for the stations' different altitudes before the charts are plotted. Isobars are normally drawn at intervals of four hectopascals (= 4 mb).

There are six main features found on such charts:

- highs (anticyclones) – areas of high pressure surrounded closed isobars
- lows (depressions, known technically as cyclones) – areas of low pressure, surrounded by closed isobars
- ridges – approximately 'V'-shaped, or elongated extensions of high-pressure areas
- troughs – extensions of low pressure areas
- cols – areas lying between approximately symmetrically placed pairs of highs and lows
- fronts

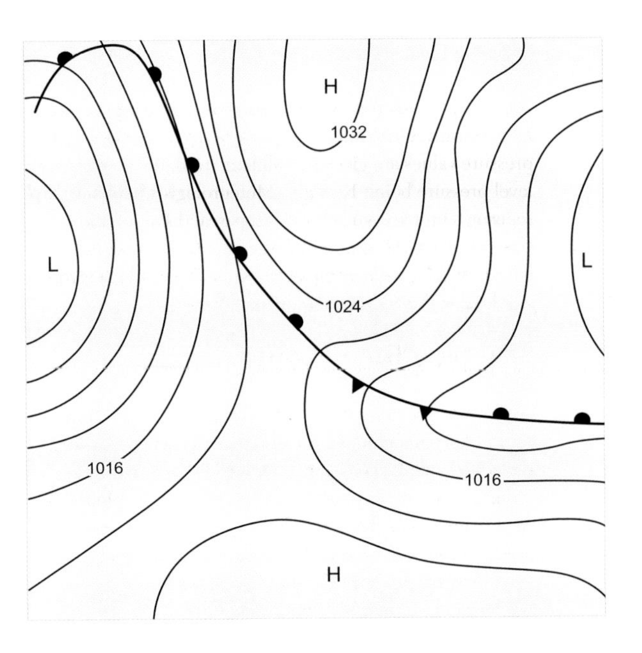

Fronts are shown conventionally by specific symbols: lines with semicircles for warm fronts, triangles for cold fronts, and both for occluded fronts (described later). The symbols are on the side towards which the front is advancing. (The symbols alternate sides for a stationary front.)

The diagram shows a col, surrounded by two highs (H) and two lows (L). Cols tend to be short-lived, because any change, however small, in the pressure pattern usually leads to a major change in the position of the col, or its complete disappearance.

Pressure systems

We have already discussed the semi-permanent high- and low-pressure regions that form part of the overall global circulation (p.186), but the weather in the temperate zones is dominated by more mobile high- and low-pressure systems. The areas of high pressure (anticyclones) tend to move fairly slowly, gradually expanding and contracting, covering larger or smaller areas with their influence, and sometimes disappearing completely. Low-pressure systems (depressions) move more rapidly, generally driven by the prevailing westerlies, but often steered north or south around the slower anticyclones.

Winds around pressure centres

As we have seen (p.186) winds are set in motion by the pressure gradient between high and low pressure regions, but at the surface their directions are modified by the Coriolis force and friction. As a result, surface winds spiral out from the centres of highs (anticyclones): clockwise in the northern hemisphere, and anticlockwise in the southern. Similarly, the surface winds spiral anticlockwise into the centres of lows (depressions) in the northern hemisphere, and clockwise in the southern. Away from the surface, the tendency is for winds to be geostrophic (p.190) and flow parallel to the isobars.

Because there is vertical motion in the centres of both anticyclones and depressions, there is an inflow and an outflow, respectively, at height, so each feature is actually accompanied by a low or a high in the upper troposphere. The air at a high level that is flowing in towards an anticyclone is also descending and warming, so no high clouds are formed. This is in contrast to the air ascending over a low-pressure region, which may cool to below

freezing, and often produces high-level cirrus clouds. This is particularly noticeable with hurricanes, where (in the northern hemisphere) satellite images often show a vast shield of cirrus spiralling out clockwise from the centre.

The pressure centres in the upper atmosphere do not precisely correspond in position, size or shape with the highs and lows at the surface. The upper-level pressure patterns and the associated jet streams have an extremely significant role in the growth, persistence, decay and motion of surface pressure systems. For this reason the numerical methods of weather forecasting used by meteorological services employ computer models that incorporate data for numerous levels in the atmosphere. Some of the most complex models use as many as 50 levels, and producing forecasts requires some of the most powerful supercomputers in existence.

Anticyclones

Although anticyclonic conditions may arise independently of the semi-permanent high-pressure zones over the middle latitudes, such weather often originates when a ridge of high pressure extends over a particular region. Such a situation, for example, often arises in summer when the Azores High extends a ridge northeast across the Iberian Peninsula, France, and other areas of western Europe.

High-pressure systems are generally accompanied by settled weather with no dramatic changes, especially as anticyclones tend to move slowly. The air is subsiding, so cloud cover is generally light or may be completely absent. In summer, the clear skies produce strong heating of the ground, but usually any convection is limited by an inversion created by the subsiding air. Thermals rising from the surface may spread out to give shallow altocumulus cloud, for example. There may be some thin high cirrus, but this is often cirrus intortus, with no obvious signs of organization by winds or wind shear. At night, however, the clear skies may lead to a considerable drop in temperature. Precipitation is usually extremely light – occasionally leading to major droughts – although sometimes the anticyclonic conditions finally break down with violent thunderstorms.

In winter, with generally colder and more humid conditions, there may be persistent stratus or stratocumulus cloud, giving rise to what is known as 'anticyclonic gloom'. When there are clear skies or very light cloud cover, low temperatures prevail at night, and radiation fog commonly forms, particularly in the autumn and early winter.

Although all anticyclones tend to move slowly, occasionally a particularly deep one extends throughout the troposphere and remains stationary for a long period of time, sometimes weeks on end. Such blocking anticyclones have a significant

effect on the weather. They force depressions approaching from the west to travel round them, either to the north or south, which may result in uncharacteristic weather for the particular season in certain regions. A blocking anticyclone over Scandinavia in winter, for example, often draws frigid air from Russia and the Siberian High over the whole of western Europe.

Depressions

As we have seen, a depression originates as a wave on the Polar Front, with the development of distinct warm and cold fronts. As the wave grows, it reaches a point where a closed wind circulation is established and a low has been created. Air converges on the centre and escapes by rising and flowing outwards at a higher level.

Although all depressions are different, they share certain common features. There is a warm front that makes a fairly shallow angle with the surface, and advances slowly, forcing the cooler air ahead of it to retreat. The cold front, by contrast, normally advances more rapidly, undercutting the warm air ahead of it. Although a front appears as a distinct line on a chart, in reality there is a zone where there is some mixing of the two air masses. This zone may be as much as 100 km wide, so it takes some time to pass over any observer.

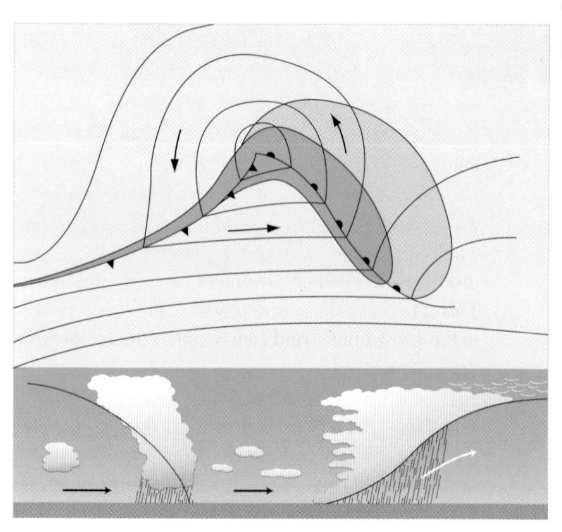

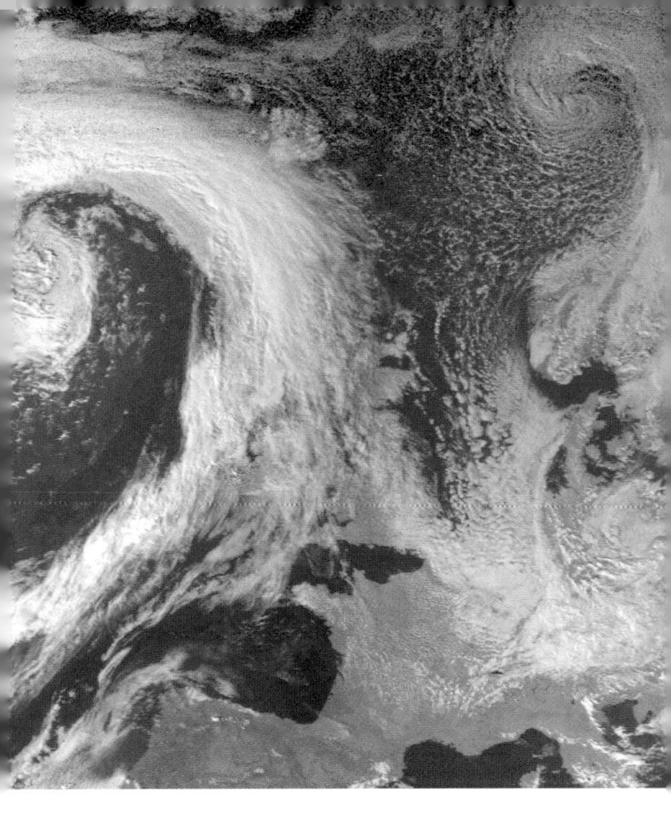

Between the two fronts lies the warm sector, consisting of maritime air. Depressions are sometimes known as 'warm-sector depressions' to distinguish them from other lows where the fronts and warm sector are absent. The isobars within the warm sector give a good indication of the probable direction of movement of the centre of the low. This is because the centre tends to move on a path parallel to the geostrophic wind (see p.190) – the upper wind – in that sector.

The lower panel of the diagram shows an idealised cross-section of a 'classic' depression along a line crossing the warm sector. There are certain distinct signs and changes that occur as a depression approaches and passes overhead.

Ahead of the warm front

Ahead of a depression, the air is generally polar in nature with moderately good visibility and fairly clear skies with some cumuliform cloud. With the approach of the warm front we find:

- wind increases and backs (n. hemisphere) towards the south; there may be signs of crossed winds (e.g., lower wind from SE, upper wind from SW)
- pressure drops, at first slowly, then more rapidly
- cloud cover increases
- precipitation is initially absent but increases, eventually becoming more or less continuous
- visibility is initially good, but then deteriorates
- temperature decreases
- dewpoint remains steady

A distinct succession of clouds is generally encountered:

- wisps of cirrus gradually increase from one direction
- the cirrus thickens to become cirrostratus
- the cirrostratus thickens and lowers to become altostratus
- the altostratus thickens and lowers, becoming nimbostratus

As mentioned earlier, warm fronts have a slope of 1:100 to 1:150. Cirrus forms at an altitude of approximately 6 km, so when the first cirrus is overhead, the warm front will be approximately 600–900 km away.

With the nimbostratus, of course, persistent rain arrives. There is generally an area of moderate rain well ahead of the surface location of the warm front, and an area of heavier rain closer to it, and following behind. These areas are shown by the different tints on the diagram (p.212). The rain is usually organised into bands of heavier and lighter rain, lying roughly parallel to the front.

The warm front arrives

With the passage of the frontal zone, a number of distinct changes occur:

- wind veers (n. hemisphere) from SE to S or S to SW
- pressure steadies
- cloud (nimbostratus) may break, depending on distance from low's centre
- precipitation generally declines (unless close to centre) and may become drizzle; farther away from centre generally ceases
- visibility becomes poor
- temperature is cool in rain, may rise if cloud breaks
- dewpoint (i.e., humidity) rises in warm air

The warm sector

In the warm sector, the cloud cover depends greatly on the distance from the centre. Close to the low's centre it may remain unbroken nimbostratus, stratus or heavy stratocumulus. Farther away it may break up completely, and there may even be sufficient convection to give cumulus clouds. Remnants of higher clouds often trail behind the warm front, so mixed skies may result.

- wind direction and strength remains steady (but strength may increase if low is deepening)
- pressure often remains steady, but may drop if low is deepening
- precipitation is light rain or drizzle, or may cease
- visibility is moderate to poor with possibility of fog at sea
- temperature generally rises
- dewpoint remains steady

As the cold front approaches

Because the slope of a cold front is away from an observer, its approach is less distinctive, and is often masked by cloud. Just ahead of the front:

- wind may back slightly and increase in strength
- pressure falls very close to front
- cloud (stratocumulus, stratus) often thickens to nimbostratus
- precipitation may occur as bands parallel to approaching front
- visibility remains moderate or poor, with possible sea fog
- temperature normally remains steady
- dewpoint also remains steady

The cold front arrives

The most dramatic changes occur when a cold front arrives:

- wind suddenly veers sharply and may become very blustery
- pressure rises abruptly
- cloud cover is often nimbostratus, but there may be significant convective activity
- precipitation is heavy, often with hail or thunder from cumulonimbus
- visibility is poor in rain showers
- temperature drops abruptly
- dewpoint falls abruptly in cold air

Because the slope of a cold front is less than that of a warm front (1:50 to 1:75), the frontal zone is generally narrower. The clouds may be similar to those at the warm front, but in reverse order. Frequently, however, there is greater instability, and embeded cumulonimbus, or even a front that largely consists of cumulonumbus, is quite common.

Behind the cold front

Finally as the cold front passes, we have:

- wind may back, then steady, generally becoming gusty and stronger
- pressure rises rapidly at first then gradually steadies
- cloud cover clears immediately behind front, then shower activity sets in
- precipitation may stop immediately behind front for 1–2 hours, then become showery
- visibility becomes extremely good in clear polar air
- temperature steadies
- dewpoint remains steady

Showers commonly occur in the air behind the cold front, and are distinctly visible on satellite images by their speckled appearance.

Occluded fronts

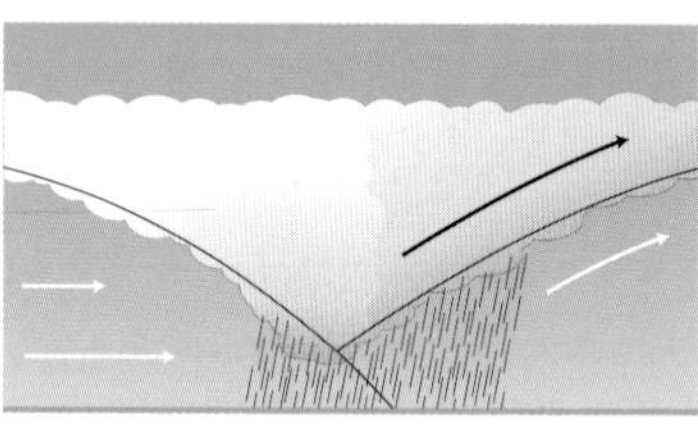

Because the cold front is faster moving, it gradually overtakes the warm front and begins to lift the warm air away from the surface. The front is now said to be occluded, and its exact characteristics depend upon whether the coldest air was ahead of the original warm front, or is behind the cold front. The first gives rise to a warm occlusion, and the second to a cold occlusion. Both are shown in the diagram, and it will be seen that although many features are similar, cold occlusions tend to have a narrower rainbelt and arrive more abruptly. A mature depression often has a long occluded front, which (as the accompanying satellite images show) may trail far behind the 'triple point' where the three types of front meet. Such occluded fronts may produce heavy, persistent rain and have often led to extensive flooding. Eventually, the frontal systems disappear as the low fills and fades away.

Weak frontal systems

On many occasions, and especially in maritime regions, the air in the warm sector tends to be stable, and there is often a slow descent of air at middle levels, leading to subdued, ill-defined frontal systems, with light precipitation. The cloud succession tends to be:

- cumulus becomes stratocumulus ahead of the warm front
- thick stratocumulus may give light rain at the front
- mixed stratocumulus and stratus with little precipitation in the warm sector
- thickening stratocumulus at the cold front
- cumulus or cumulonimbus behind the cold front.

Polar lows

Occasionally a weak secondary depression without fronts forms in the unstable air behind the main cold front of a depression. It normally consists of a line of significant convective activity and the associated precipitation along a trough that extends away from the primary depression. Such a trough line is shown on the chart, which also illustrates how secondary depressions normally begin as waves on the trailing cold front, and closer to the equator than the primary depression.

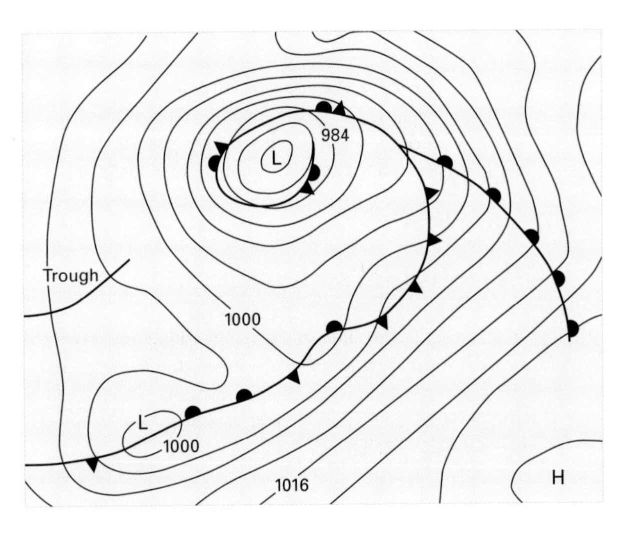

Showers

ID FACT FILE

Occurrence:
In a cold air stream passing over a warm surface (such as the sea or large lakes), typically after the passage of a cold front; after strong convection over the land during the day.

See also:
cumulus congestus (p.55); cumulonimbus (p.74); cold and warm fronts (p.104); hail (p.178); rain (p172); snow (p.176)

In meteorology, the term 'shower' has a very specific meaning. It implies that cumulus congestus or cumulonimbus clouds are producing fairly short, but relatively heavy periods of rain. This is in contrast to the extended period of frontal rain, punctuated by heavier bursts, that is commonly found at warm or occluded fronts in depressions. Cold fronts often have cumulus congestus and cumulonimbus embedded within them, and showery weather frequently occurs after the passage of a cold front.

In winter, both the condensation and freezing levels are low, and convective clouds are fairly shallow. Precipitation is in the form of snowflakes or ice crystals, although these may melt to rain if there is a warmer layer near the surface. Rain or snowfall is generally relatively short, unless the clouds are downwind of a major source of warm water, such as the sea or

the Great Lakes in North America, when a succession of showers may produce prolonged heavy precipitation. In early winter, cold air flowing out of Canada, for example, crosses Lake Erie, which is still unfrozen, and picks up large quantities of moisture, which gives rise to extremely heavy snowfall over Buffalo, New York State, and the surrounding area.

In summer, temperatures are higher, as are the condensation and freezing levels. The moisture content of cumulus congestus and cumulonimbus clouds is much greater, and they are much deeper. Cumulus congestus

may produce heavy showers by the 'warm rain' process, and cumulonimbus create very heavy precipitation in the form of rain or hail. In cumulonimbus, in particular, the strong downdraughts may produce sudden bursts of heavy precipitation.

The lifetime of showers is fairly limited. They go through three stages:

- early stage: The cloud is cumulus congestus, actively growing, sometimes with large raindrops
- mature stage: The cloud may turn into cumulonimbus calvus and capillatus. Most precipitation occurs at this stage, with large raindrops (and possibly hail); there is vigorous convection
- late stage: Downdraughts swamp the updraughts and rainfall decreases in intensity and size of raindrops

The first two stages each last about 20 minutes, but the final stage may last between 30 minutes and about 2 hours.

The downdraughts, apart from eventually quenching the updraughts, spread out when they reach the surface, producing gusts of wind directed away from the cloud itself. It is this wind that has resulted in the mistaken belief that showers may approach 'against the wind'. In fact, showers generally follow the (geostrophic) wind at low cloud height.

Storms

ID FACT FILE

See also:
cumulonimbus (p.74); hail (p.178); lightning (p.226); showers (p.220)

When conditions are suitable, cumulonimbus clouds may develop into major storms. The cold downdraughts from one cell, for example, may spread out ahead of the cloud, undercut warm air, and initiate another cell some distance away from the parent storm. Frequently, too – even in quite small shower systems – air flowing into a strong cumulonimbus creates a line (known as a flanking line) of daughter cells, which increase in size and intensity towards the parent system. When more than one cell is producing rain, hail, or lightning at any one time, the system has become a multicell storm.

As described earlier, each cell goes through growing, mature, and decaying stages, so in a

multicell storm there are cells in various stages of development. As one cells decays, another takes over activity. This is particularly noticeable in thunderstorms (especially at night), where the lightning from one cell gradually ceases, and the main centre of activity shifts to another part of the storm. In major storms several cells may be active at once.

Multicell storms often become organized into a long line of cells. New cells arise in front of the line, where humid air is being drawn in, so the whole line (known as a squall line) becomes self-perpetuating and travels across country, lasting for many hours.

In extreme cases, the exceptionally strong updraughts and downdraughts may become organized into a single giant cell – a supercell – generally with one giant updraught 3–10 km across. Such supercells are (like multicell storms) self-propagating, and they may produce exceptionally heavy rain and violent hail, as well as spawning tornadoes. They may last for hours or even days if conditions are suitable.

Lightning

ID FACT FILE

SEE ALSO: *cumulonimbus (p.74); showers (p.270); multicell and supercell storms (p.224)*

Lightning is an electrical discharge between a cloud and the ground, between clouds, or between different regions of the same cloud. When the discharge channel is visible, it is commonly called 'fork lightning' and when invisible, being masked by cloud, 'sheet lightning', but these terms have little true significance. Similarly, when lightning is too distant for the thunder to be heard, it is sometimes called 'heat lightning'.

Lighting may occur in individual isolated cumulonimbus cells, in multiple cells, or in a supercell. Except for the last, individual cells are usually active for no more than 30–45 minutes. If the bearing to an active cell remains the same, then you can assume that it will pass overhead.

The exact mechanism by which positive and negative charges are separated remain uncertain, but it is believed that strong updraughts lift positively charged particles to the top of the cumulonimbus, while heavier, negatively charged particles fall to the bottom. The ground beneath the cloud becomes positively charged. A flash occurs when the electrical resistance of the air breaks down and the initial discharge (the stepped leader) creates a channel between cloud and the ground. The main discharge current then flows upwards into the cloud. The process may recur several times in a single flash.

The intense heating by a flash causes the air to expand and then contract violently, producing thunder. The number of seconds between flash and thunder divided by three gives the approximate distance of the lightning in kilometres.

Lightning safety

Indoors: stay away from windows, water pipes, and telephone and electrical wiring. If possible, disconnect phones, computers, etc. Outdoors: Avoid open spaces where you are the highest object. Take shelter in a building or car. Do not stand under isolated trees. If no cover, choose the lowest spot (a valley rather than hills), or a dry ditch. Do not lie flat, instead crouch with arms around legs and head on knees. Boats are very vulnerable: on small sailing boats with wire shrouds, wrap anchor chain round a shroud and dangle one end in the water to provide an earth channel.

Devils and whirls

ID FACT FILE

APPEARANCE:
A rotating column of air, revealed by the material that it raises from the surface (such as dust, snow, or water).

OCCURRENCE:
Where the funnelling effect of wind, or uneven heating or friction of the surface initiates rotation.

SEE ALSO:
tuba (p.86); tornadoes (p.232); landspouts and waterspouts (p.230)

Any rotating column of air that originates at the surface and extends upwards is known as a 'devil' or, more formally, as a 'whirl'. In this respect they are unlike funnel clouds (tubas), landspouts, waterspouts, or tornadoes, which grow downwards from higher cloud until they reach the ground. In addition, unlike the last three, they rarely cause any damage.

A column of air may be set into rotation when the wind funnels through a narrow gap between hills or mountains. As it spins it generally raises material from the surface, and is usually named after the type of material, so there are dust, water, and snow devils (or whirls). In cities, similar rotating columns of paper and litter are often created by winds gusting between high buildings. Most such whirls remain more-or-less stationary until wind conditions change.

Uneven heating of the ground and variations in its roughness may also produce whirls, and dust devils are very common in dry, hot, desert areas, where frequently more than one may be seen at the same time. Such devils often move away from their place of origin, unlike the whirls created by funnelling effects.

Such devils originate when intense heating of the ground produces a thin, extremely hot layer of air next to the surface. This in turn creates a small, but very strong updraught,

which drags in more hot air at the lowest level. The size of dust devils is generally too small for them to be affected by the Coriolis effect (p.187) and their initial rotation is often influenced by variations in friction at the surface.

Powerful fires sometimes create similar 'fire devils' and these often become extremely strong, because the inflow of air provides plenty of oxygen to feed the flames. The concentrated updraught often carries burning fragments which it scatters over the surroundings, frequently leading to a number of spot fires around the main fire and thus actively helping the fire to spread.

Landspouts and waterspouts

ID FACT FILE

APPEARANCE:
A grey or white funnel cloud that reaches the surface, and may raise water, dust, hay or other materials.

OCCURRENCE:
Beneath cumulus congestus or cumulonimbus clouds with strong downdraughts.

SEE ALSO:
cumulus congestus (p.32); cumulonimbus (p.74); tornadoes (p.234); tuba (p.86)

The term 'landspout' is of fairly recent origin and arises because both they and waterspouts are now known to form by the same process, which differs greatly from the mechanism that is responsible for tornadoes. A strong downdraught falls from the bottom of a cumulus congestus or cumulonimbus cloud. Reduced pressure within the downdraught causes condensation, producing a funnel cloud (tuba), surrounded by a rising column of air, which is normally invisible. The funnel may appear grey or white.

When the funnel reaches the surface, it becomes a landspout or waterspout. In the latter, the flattened area where the tip touches the water is known as the 'dark spot', which is usually surrounded by a curtain of spray, known as the 'bush'. Landspouts also frequently show a short column of material such as dust, hay, or other loose material that has been lifted from the ground.

Landspouts and waterspouts are far less destructive than tornadoes, and at least one sailor has deliberately sailed into a waterspout – 'to see what it was like'. Such foolhardy behaviour is not to be encouraged, however, because both forms may cause considerable damage to weak structures and vessels.

Tornadoes

ID FACT FILE

Appearance:
A funnel cloud that reaches the surface and raises an outer column of debris.

Occurrence:
Beneath a supercell system, which is normally accompanied by heavy rain and hail, strong gusts, and thunder and lightning.

See also:
landspouts and waterspouts (p.230); severe storms (p.225); tuba (p.86)

Tornadoes are generated by a different mechanism from the one that creates landspouts and waterspouts. They are also far more destructive, but luckily are much rarer. (Most so-called tornadoes mentioned in the news media are actually funnel clouds, landspouts or waterspouts.) Most tornadoes occur in the United States, where the large land-mass and great temperature contrasts between air masses favour their formation.

True tornadoes arise from powerful supercell storms. There is a large-scale rotation of air at middle levels within the storm, which produces violent downdraughts and updraughts. An area of cloud that is normally rain-free descends from the main storm, and often develops into a rotating cylinder of cloud – the 'wall cloud'. This is a sign that a tornado is probably imminent. The tornado itself descends either from the wall cloud or from a nearby rain-free cloud-base. Initially the funnel may be invisible, but as soon as it touches the ground it raises a cloud of debris from the surface (and then is officially known as a tornado).

Tornadoes generally have wind-speeds of 150–300 kph – 512 kph is the record – which is why they are so destructive. They may move across country, with paths that are normally 10–100 km long. Official forecasts issue two stages of alert:

- watch: conditions are appropriate for the formation of tornadoes. Take precautions and make sure you hear further notices.
- warning: a tornado is imminent or has been seen in your vicinity. Take cover immediately in a tornado shelter.

Tropical cyclones

ID FACT FILE

WIND SPEEDS:
Over 120 kph

ORIGIN:
5–10° N or S of equator, where sea surface temperatures exceed 27°C

LARGEST CLOUD SHIELD:
3500 km in diameter for Hurricane Gilbert over Caribbean in 1988

LOWEST PRESSURE:
870 mb in typhoon Tip, west of Guam in the Pacific Ocean, 12 October 1979

HIGHEST STORM SURGE:
7.3 metres for Hurricane Camille at Pass Christian, Mississippi, 7 August 1969

Tropical cyclones are known by various names: 'hurricane' in the Atlantic and eastern Pacific Oceans, 'typhoon' in the northern Pacific Ocean, and 'cyclone' in the Indian and western South Pacific Oceans. All are large circulating storm systems with extremely high wind speeds. They consist of a low-pressure core, surrounded by spiral bands of extremely vigorous convection. The cloud systems may be thousands of kilometres across.

The devastating effects of tropical cyclones arise from three main factors. They have extremely high sustained wind speeds and, in addition, they often spawn even stronger small tornadoes. The low pressure in the centre and the winds create a dome of water, sometimes several metres higher than normal sea levels. Such storm surges normally cause most of the devastation when hurricanes hit the coast, and have been known to travel more than 15 km inland.

The torrential rain often causes flooding and triggers landslides and mudslides, which may sweep away roads, bridges, railways, and whole towns. In some countries, however, agriculture is dependent on the rains that accompany tropical cyclones, so they do have some beneficial effects.

With modern satellite imagery and fairly sophisticated computer models, it is generally

possible for forecasters to predict the track of tropical cyclones with a considerable degree of accuracy. It is not yet possible, however, to predict wind speeds or the amount of precipitation with a great deal of confidence.

Tropical cyclones die away when they pass onto land, or when they reach cold waters. Frequently, however, they reach middle latitudes and merge with pre-existing depressions, causing them to deepen and become extremely vigorous.

Weather satellites

Weather satellites have become an indispensable part of modern weather forecasting. Their instruments provide meteorologists with a vast amount of data covering the whole globe. Satellites have the great advantage that they give complete coverage of the oceans and other areas that are too remote to have surface observing stations.

The information provided by dedicated meteorological satellites is supplemented by observations from many other satellites whose primary purpose is to monitor aspects of the Earth's environment. Such data is not intended (or used) for day-to-day forecasting purposes, but may be employed for climatic research or the study of specific aspects of the weather, such as the development and evolution of tropical cyclones.

Geostationary satellites

The general public is most familiar with the images that are often included – sometimes as time-lapse sequences – in television weather forecasts. Two forms of image are encountered, each provided by a different type of satellite. The time-lapse sequences are usually obtained from images obtained at fixed intervals by geostationary satellites. These orbit 35,900 km above the equator, where their orbital period is 24 hours, and they remain stationary above one point. Complete coverage of the tropics and temperate regions is ensured by the use of five or six satellites spaced around the equator, but the polar regions are poorly covered because they are masked from the satellites' view by the curvature of the Earth.

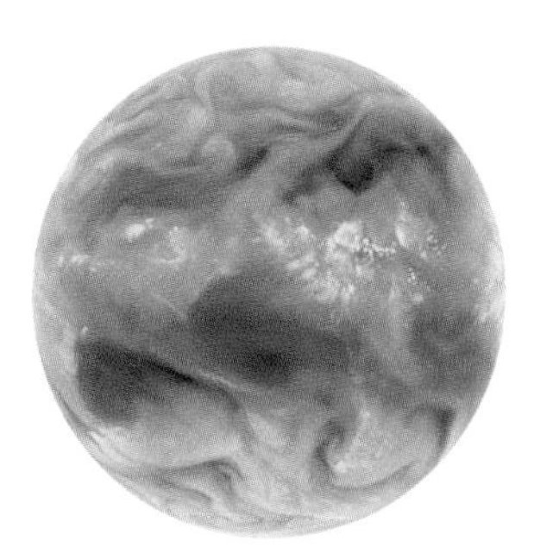

Geostationary satellites' big advantage is that they offer essentially continuous coverage of the area beneath them, with the newer satellites returning images taken at several wavelengths every 15 minutes. Although the most recognizable images are those taken at visible wavelengths (top *left*), others include images in the infrared, obtained both by day and night (*centre*), and striking images showing the water-vapour content of the atmosphere (*bottom*).

Geostationary images – winter

These two images were obtained on 24 February, just after the coldest period of the northern winter. An interesting feature is the Intertropical Convergence Zone (ITCZ), marked by the distinct line of clouds running across the Atlantic, and which always remains north of the equator in this part of the world. A cloud-free, high-pressure area lies over

North Africa and Arabia, and a much smaller one is visible over Southern Africa. There are three distinct depressions: two (over the North and South Atlantic) have distinct bent-back occluded fronts, and show the lines and speckled areas behind their cold fronts where convective showers have developed over the relatively warm oceans. The depression over the East Coast of North America, by contrast, has a large, cloud-free area behind the cold front, where a surge of cold air has moved south from Canada.

Geostationary images – Summer

This pair of images were obtained on 30 August, both at approximately local noon, but the left-hand (western) image was about five hours after the right-hand one. Changes in the clusters of convective clouds off the west-African coast are visible. Such clusters (usually thunderstorm centres) are particularly marked across equatorial Africa, and those that move

west across the Atlantic sometimes give birth to hurricanes, one of which (Emily) is clearly visible in the north Atlantic. Other strong convective clusters are visible in the western hemisphere. There is a series of weak depressions at moderately high northern latitudes, but none are visible in the southern hemisphere. Although the high-pressure zone persists over north Africa and Arabia, the western Sahara and western coast of Arabia have more cloud than in the wintertime image.

Polar-orbiting satellites

The big disadvantage of geostationary satellites – that they do not provide good coverage of high latitudes – is met by the polar orbiting satellites. These operate at a much lower altitude (800–1000 km) than the geostationary satellites, and have orbital periods that are about 90–100 minutes. These satellites continuously scan a swathe of the surface beneath them. The orbit remains essentially fixed in space and as the Earth rotates beneath it, the satellite scans an adjacent swathe at each subsequent orbit.

Although, being at lower altitudes, they have greater resolution than geostationary satellites, their main disadvantage is that generally only a few passes per day cover any particular location. So they do not provide continuous coverage of any one area, but they do return images from the whole Earth, and thus complement the geostationary data.

As with the geostationary satellites, data is obtained at several different wavelengths, including visible and infrared channels.

Although images in the visible channel (*left*) are the most familiar, infrared images (*below*) are extremely informative and generally show the coldest clouds as white, and warm areas of the surface (land or sea) as dark. Images at different wavelengths are often combined to give false, but reasonably realistic, colour images (*bottom*).

Clouds and weather

Forecasting the weather is a highly complex process, and professional meteorologists use data from across the whole world, and the largest and fastest supercomputers to produce their forecasts. Even so, local circumstances, such as the observer's altitude, the presence of hills and mountains, or proximity to the sea may alter the conditions that are actually experienced in any given area.

Clouds	Page	Description
cumulus humilis	28	Scattered and moderate in size
cumulus humilis and cirrus	214	Cumulus fail to grow, cirrus increases
cumulus congestus	32	Towering cumulus
cumulus spreading into stratocumulus or altocumulus	41, 46	Decreasing clear sky, inversion limiting growth
cirrus	58	High thin clouds, few in number
cirrus thickening to cirrostratus	70, 214	Halo may appear around the Sun or Moo
cirrus or cirrostratus and altostratus	52, 214	Cloud obviously increasing; strong upper wind spreads any contrails
jet-stream cirrus	114	High, fast-moving bands of cirrus often with wave structure
contrails	108	Disappear rapidly

Some indication of the likely development may be gained from observation of the sky, clouds, and winds, especially if it is possible to look at the sky several times a day to follow changes as they take place. Remember that, in general, at temperate latitudes depressions tend to move from west to east, although there are exceptions, such as the northeasters – the cyclonic storms that affect New England in winter and often bring major snowfall.

Forthcoming weather
Fair weather, unless deeper cumulus congestus, high cirrus or cirrostratus appear
Warm front approaching, especially if cirrus turns into cirrostratus; rain probably within 12 hours
Showers are possible within the next few hours
Precipitation unlikely, unless convection strong enough to break through inversion; overcast likely to be slow to clear
Generally fair, but weather may deteriorate if clouds increase and become organized, spreading from one particular direction
A frontal system is approaching with deteriorating weather; rain likely within 6-12 hours
Warm front approaching, wind likely to increase, rain within 6–12 hours
A vigorous depression exists upwind and is approaching; cloud cover will increase; and strong winds are likely in 10–15 hours
Fine, major change unlikely within 24 hours, unless conditions conducive to major convective activity

Clouds	Page	Description
contrails spreading	110	Trails spread, become glaciated and persistent
altocumulus floccus or castellanus	48, 49	Clumps of altocumulus (often with virga) or lines of towers
altocumulus and cirrocumulus	46, 66	Increasing in extent, becoming altostratus and cirrostratus
altostratus and pannus	52	Sky covered with altostratus, ragged pannus increasing
nimbostratus	56	Sky completely overcast
cumulonimbus	74, 220	Individual clouds of limited horizontal extent
cumulonimbus	224	Large clouds with several cells and/or anvils
cumulonimbus	224	Organized lines or clusters of massive clouds
stratocumulus	40	Low cloud with occasional breaks
mixed, stable clouds	215	Stratocumulus, altostratus and cirrostratus patches
stratus	34	Low cloud

Some of the factors that may modify the general forecast and which depend on local conditions are:

- hills or mountains – p.94
- rain shadow – p.95
- lakes – p.200 & p.221
- sea breeze (and land breeze at night) – p.196
- valley wind (and mountain wind at night) – p.194

Forthcoming weather
Increasing humidity at height; possible indication of approaching warm front with deteriorating weather
Thundery showers (possibly severe) likely within 24 hours
Rapidly deteriorating weather, rain within 6–12 hours
Warm front nearby; rain imminent, starting intermittently, but becoming essentially continuous; wind likely to increase
More-or-less continuous rain (usually bands of heavy rain interspersed with periods of lessened or no rain), lasting for several hours
Light to moderate showers, gusts; active lifetime limited to 20–30 minutes
Heavy showers with possibility of hail and lightning; strong gusts; lifetime perhaps as long as a few hours
Extremely heavy showers, possibly beginning with severe hail; if part of a cold front will be followed by scattered, and possibly heavy, showers
No significant precipitation; likely to be slow to clear
Warm sector cloud; no significant change imminent and may persist for days
No significant precipitation; if originated as fog, will probably clear later in day, otherwise persistent

- föhn effects – p.200
- fall winds – p.201

All of these different factors can significantly alter the wind direction and strength, the amount and type of precipitation, the degree of cloud cover, and the air temperature.

GLOSSARY

anticyclone A high-pressure region where air subsides from higher altitudes and flows out over the surrounding area. The circulation around an anticyclone is clockwise in the northern hemisphere.

antisolar point The point on the sky directly opposite the position of the Sun.

backing An anticlockwise change in the wind direction, e.g. from West, through South, to East.

col An area of slack atmospheric pressure, lying between a pair of low-pressure centres and a pair of high-pressure ones. Slight changes in the pressure pattern may cause a col to move rapidly or disappear abruptly.

continental climate A climate found in continental interiors, characterized by extremely cold winters and hot summers. Generally has low overall precipitation.

convection Transfer of heat by the movement of a fluid such as air or water.

Coriolis force The apparent force that deflects any moving object (such as the wind) away from a straight-line path. In the northern hemisphere it acts towards the right, and in the southern, to the left. The greater the speed of the moving object, the greater the Coriolis force.

cyclone A weather system where air circulates around a low-pressure core, with two distinct meanings: 1) a 'tropical cyclone', a self-sustaining tropical storm, also known as a hurricane or typhoon; 2) an 'extratropical cyclone' or depression, a low-pressure area, one of the main weather systems in temperate regions.

cyclonic Moving or curving in the same direction as air flows around a cyclone, i.e., anticlockwise in the northern hemisphere, clockwise in the southern.

depression A low-pressure area, known technically as an 'extratropical cyclone'. Air flows into a depresssion and rises at its centre. The wind circulation around a depression is cyclonic (anticlockwise in the northern hemisphere).

dewpoint The temperature at which a particular parcel of air, with a specific humidity, reaches saturation. At the dewpoint, water vapour condenses into droplets, producing a cloud, mist or fog, or depositing dew on the ground.

hurricane The name applied to a tropical cyclone that occurs in the North Atlantic or eastern Pacific.

instability The condition under which a parcel of air, if displaced upwards or downwards, continues (or even accelerates) its motion. The opposite is stability.

inversion An atmospheric layer in which temperature increases with height.

isobar A line on a weather chart that joins points with the same barometric pressure, generally adjusted to give the reading that would apply at sea level.

jet stream A narrow ribbon of high-speed winds that lies close to a break in the level of the tropopause, with two main jet streams (the polar-front and sub-tropical jet streams) in each hemisphere. Other jet streams exist in the tropics and at higher altitudes.

lapse rate The rate at which temperature changes with increasing height. By convention, the lapse rate is positive when the temperature decreases with height, and negative when it increases.

latent heat The heat that is released when water vapour condenses, or freezes into ice. It is equivalent to the heat that was originally required for evaporation or melting.

maritime climate A climatic regime that is strongly influenced by a nearby ocean. Generally characterized by significant amounts of precipitation throughout the year, but normally with mild winters and summers without extremely high temperatures.

mesosphere The atmospheric layer above the stratosphere, in which temperature decreases with height, reaching the atmospheric minimum at the mesopause, at an altitude of either 86 or 100 km (depending on season and latitude).

parhelion The technical term for a mock sun.

precipitation The technical term for water in any liquid or solid form that falls from clouds and reaches the ground. It excludes cloud droplets, mist, fog, dew, frost and rime, as well as virga.

pressure tendency The change in atmospheric pressure during the previous three hours.

rain shadow An area of lesser overall rainfall to leeward of high ground.

ridge An extension of an area of high pressure, resulting in approximately 'V'-shaped isobars pointing away from the pressure centre.

stratosphere The second major atmospheric layer from the ground, in which temperature initially remains constant, but then increases with height. It lies between the troposphere and the mesosphere, with lower and upper boundaries of approximately 8–20 km (depending on latitude) and 50 km, respectively.

stability The condition under which a parcel of air, if displaced upwards or downwards, tends to return to its original position rather than continuing its motion.

supercooling The conditions under which water exists in a liquid state, despite being at a temperature below 0°C. This occurs frequently in the atmosphere, often in the absence of suitable freezing nuclei.

synoptic chart A chart showing the values of a given property (such as temperature, pressure, humidity, etc.) prevailing at different observing sites at a specific time.

thermal A rising bubble of air, which has broken away from the heated surface of the ground. Depending on circumstances, a thermal may rise until it reaches the condensation level, at which its water vapour will condense into droplets, giving rise to a cloud.

tropopause The inversion that separates the troposphere from the overlying stratosphere. Its altitude varies from approximately 8 km at the poles to 18–20 km over the equator.

troposphere The lowest region of the atmosphere in which most of the weather and clouds occur. Within it, there is an overall decline in temperature with height.

trough An elongated extension of an area of low pressure, which results in a set of approximately 'V'-shaped isobars, pointing away from the centre of the low.

veering A clockwise change in the wind direction, e.g. from East, through South, to West.

wind shear A change in wind direction or strength with a change of position. Vertical wind shear exists if the wind strength alters with a change in height. Horizontal wind shear exists if the wind strength alters with a change in position at a specific level.

zenith The point on the sky directly above the observer's head.

Further reading

Brettle, M. & Smith, B. (1999), *Weather to Sail*, Crowood Press
Chaboud, R. (1996), *How Weather Works*, Thames & Hudson
Dunlop, S. (1999), *Collins Gem Weather*, HarperCollins
Dunlop, S. (2001), *Dictionary of Weather*, Oxford University Press
Dunlop, S. (2002), *How to Identify Weather*, HarperCollins
Eden, P. (1995), *Weatherwise*, Macmillan
File, D. (1996), *Weather Facts*, Oxford University Press
Harding, M. (1998), *Weather to Travel*, Tomorrow's Guides
Ludlum, D.M. (2001), *Collins Wildlife Trust Guide Weather*, HarperCollins
Pedgley, D. (1980), *Mountain Weather*, Cicerone Press
Meteorological Office (1982), *Cloud Types for Observers*, Stationery Office
Watts, A. (2000), *Instant Weather Forecasting*, Adlard Coles Nautical
Watts, A. (2001), *Instant Wind Forecasting*, Adlard Coles Nautical
Whitaker, R., ed. (1996), *Weather: The Ultimate Guide to the Elements*, HarperCollins

Magazines

Weather (monthly), Royal Meteorological Society, 104 Oxford Road,
Reading, Berks. RG1 7LJ
(http://www.royal-met-soc.org.uk/weather.html)
Weatherwise (bi-monthly), Heldref Publications,
1319 18th Street NW / Washington, D.C. 20036-1802
(http://www.weatherwise.org/)

INTERNET SOURCES (Current weather)

BBC Weather (http://www.bbc.co.uk/weather)
CNN Weather (http://www.cnn.com/WEATHER/index.html)
ITV Weather (http://www.itv-weather.co.uk)
The Weather Channel (http://www.weather.com/twc/homepage.twc)

INDEX